Angel COMMUNICATION CODE

RESPONDING TO THE EXTRATERRESTRIAL MESSAGE

STEPHEN J. SILVA

Angel Communication Code
RESPONDING TO THE EXTRATERRESTRIAL MESSAGE

iUniverse books may be ordered through booksellers or by contacting:

iUniverse
1663 Liberty Drive
Bloomington, IN 47403
www.iuniverse.com
844-349-9409

ISBN: 978-1-6632-5926-4 (sc)
ISBN: 978-1-6632-5924-0 (hc)
ISBN: 978-1-6632-5925-7 (e)

Library of Congress Control Number: 2024900093

Print information available on the last page.

iUniverse rev. date: 02/05/2024

Dedication

This project is inspired by, and dedicated to, my incredible wife Rosemarie Silva and my children Quinn and Maxwell Silva. This project is also dedicated to my mother and father (God rest their souls); my brothers, my sister, and all their sons and daughters; and the entire Silva clan. They have all given me a gift impossible to return in kind. I understand that without the inclusion of all these people in my life, I am nothing. Their presence in my life has been, is now, and continues to be an incredible gift from God.

The world of pure spirits stretches between the divine nature and the world of human beings; because divine wisdom has ordained that the higher should look after the lower, angels execute the divine plan for human salvation: they are our guardians, who free us when hindered and help to bring us home.

—Saint Thomas Aquinas

When forced to summarize the general theory of relativity in one sentence: Time and space and gravitation have no separate existence from matter. Physical objects are not in space, but these objects are spatially extended. In this way the concept "empty space" loses its meaning. The particle can only appear as a limited region in space in which the field strength or the energy density is particularly high. The free, unhampered exchange of ideas and scientific conclusions is necessary for the sound development of science, as it is in all spheres of cultural life. We must not conceal from ourselves that no improvement in the present depressing situation is possible without a severe struggle; for the handful of those who are really determined to do something is minute in comparison with the mass of the lukewarm and the misguided. Humanity is going to need a substantially new way of thinking if it is to survive!

—Albert Einstein

Contents

Preface

Unidentified aerial phenomena (UAP) is the new politically correct term for unidentified flying objects (UFO). The name change means nothing. The issue remains the same. Such is the way that science tends to deal with this subject. What truly matters is figuring out how to communicate with these entities. The answers to the communication questions lie herein.

Much has been written on the subject of hidden codes and messages lurking in the shadows out of plain sight within the Bible. Most allude to the interpretation of prophecies and predictions of the future. Books have also been written about hidden codes and messages specifically from angels. Those books inevitably tend to lead down the path of numerology and the concept of angel numbers. *Angel Communication Code* is not about any of those things; it is about presenting hard and defensible facts. It is about the discovery of a specific code or cipher provided by either angels or ETs, or both, that was left for humans to find and that leads us to a specific place for a singular purpose, which is establishing two-way extraterrestrial (ET) communication. Perhaps to say that the plan is for us to reestablish communication is more accurate.

Angel Communication Code makes many historical references. When this information is presented, it is important that you, reader, pause and think about these dates. You must truly appreciate how long ago relevant events took place and what they mean in the context of ET existence, ET communication, and our current place in time and space. It is equally important to focus your attention on the prime numbers 2, 3, 5, and 7. Make those numbers red flags in your head right now and as you read *Angel Communication Code*. It is critical that you focus on these dates and prime numbers. You will soon see why.

Angel Communication Code provides new information that supports the conclusions cited in the author's debut book, *Extraterrestrial Communication Code*. All these conclusions support the theory that an ET code has been deposited for humans to find and decipher that leads to a way to establish two-way communication between humans on Earth and intelligent beings living elsewhere in the universe. This new book points to angels and other biblical messages as being involved in the creation of that ET communication code. The belief in codes and clues regarding ETs coming to Earth and leaving messages is not a new concept. *Angel Communication Code* merely offers a different approach to an ancient problem. It is not about the question of ET existence but, rather, about establishing two-way communication with ETs, whom we know exist. The theories, discoveries, conclusions, and hypothesis, and the proposed experiment developed and presented in *Angel Communication Code*, are an unprecedented approach to our establishment of two-way ET communication. There is a difference between ET contact and intelligent two-way ET communication. In *Angel Communication Code*, we will examine the role of angels and other biblically referenced beings and their involvement in the deposition of a cipher presented by them in a variety of ways specifically for a specific purpose.

People have been searching for hidden messages from angels in scripture since the beginning of Christianity. Sometimes the search was in the form of allegorical interpretation, where the literal meanings were secondary to a much greater spiritual meaning. The first people to claim that there was a message beyond the plain words of scripture are believed to be the Gnostics of the late first century. They believed that there was a higher message in scripture that only the enlightened could understand. The term *Gnostic* comes from γνῶσις *gnōsis*, the Greek word for knowledge.[1] Those who are trying to use computers to find numeric values or letters of the Hebrew alphabet are ignoring the plain words of scripture. What is most important in the Bible has been clearly revealed in the Bible, so why go on an expedition for some sort of hidden message treasure hunt when there is no need?[2]

The reason is this: perhaps what is being overlooked is the influence of angels in our quest to establish two-way communication with ETs. Angels are the messengers of God. It is entirely possible that ETs had influence over at least some of the authors of the books of the Bible. This influence may have included the embedding of a code for communicating with ETs through the messengers we call angels, the highest form of messengers known to humans.

The universe we thought that we might be beginning to get a grasp on seems to be more unexplained every day. It seems the more we discover, the less we know. With respect to pure science, it is understood that there is much about life in the universe that we do not know. It is also understood that there is also much about life on earth that we do not know. The same is true for scripture and the workings of angels. The most learned theologians still to this day cannot agree on certain critical interpretations of the messages revealed in scripture in general and, more specifically, revealed by angels. The absolute truth in all the science and faith-based belief on this matter may never be known. Establishing the lifeline of communication with our ET brothers and sisters, however, would be a gigantic step in the right direction in the pursuit of universal truth.

Angel Communication Code will examine only a small (but relevant) portion of the entirety of existing evidence and expose new perceptions that may have been provided by angels or by those humans who defined the structure of angels. At the heart of it all is prime number 3 and the first three prime numbers in the series, 2, 3, and 5, which mysteriously, conspicuously, and vividly appeared to this author in a specific pattern that we will call a code or cipher. These sorts of visions have been experienced by many others throughout history (cited herein), leading to many important, Nobel Prize–winning scientific discoveries. *Angel Communication Code* is not born out of one man's unique psychotic episode and vivid imagination. It is about something very real. The first three prime numbers are the foundation of the code's discovery and interpretation. The journey we will travel in *Angel Communication Code* is driven by the incontestable historical, mathematical, and geographical facts presented herein.

The concepts and facts cited in *Angel Communication Code* might encroach upon traditional Christian interpretations of angels and their role in human history. As is the case with *Extraterrestrial Communication Code*, there are discoveries revealed in *Angel Communication Code* that are unique and that directly apply to this two-way ET communication project. We must acknowledge and accept that there is room for a broader interpretation of traditionally accepted Christian (and all religious) teachings and beliefs regarding intelligent ET life in the universe and the role of angels woven throughout the entire story.

---

1 Lita Sanders, "Hidden Messages in Scripture," Creation Ministries International, https://creation.com/hidden-messages-in-scripture, August 25, 2011.

2 Jack Wellman, "Are There Hidden Messages in the Bible?," Faith in the News, accessed November 22, 2023.

The intent of the presentations made in *Angel Communication Code* is not to undermine faith or beliefs regarding Christian angels or angel equivalents from other world religions. The intent is to present the applicable facts in a way that legitimizes the possibility that humans have been provided clues that point to a code that unlocks a way to establish meaningful communication with intelligent life not of this earth.

Angel Communication Code is not intended to be about faith or religion. It is more about deciphering information that the angel themselves, or the use of angels as messengers, may have provided for us to find, figure out, and use to establish communication with the larger universe.

What we have done in the modern era to communicate intelligently with ETs has not worked; therefore, we must rethink all of it and revise the experiment. There is no other viable option. Religious teachings and beliefs can and do coexist with the existence of ETs. This synergy is woven into the multidimensional fabric of the universe and all life within it.

CHAPTER 1

Science, Faith, and the Universe

What is the universe? Easy answer—the universe is everything. It includes all of space and all the matter and energy that space contains. It even includes time itself. It encompasses everything that ever was or ever will be. The universe is the sum of all that exists. It includes all matter, from a single electron on earth to everything that makes up all galaxies. The universe also includes all forms of energy. There is nothing in theory outside the universe because anything that exists is, by default, included in our current definition of the universe.

Now consider the concept of the multiverse. The multiverse is a hypothetical group of multiple universes. Together, these universes are presumed to include everything that exists: the entirety of space, time, matter, energy, information, and the physical laws and constants that describe them. The different universes within the multiverse are called "parallel universes," "other universes," "alternate universes," or "many worlds." One common assumption is that the multiverse is a "patchwork quilt of separate universes all bound by the same laws of physics." The concept of multiple universes, or a multiverse, has been discussed throughout history, with origins in ancient Greek philosophy. It has evolved over time and has been debated in various fields, including cosmology, physics, and philosophy.[1] Whatever is out there, the hard part is how to identify it, understand how it all works together, and discover how to communicate with these realms. There are many ways to communicate.

There are people out there who wholeheartedly believe that there is a thing in our universe called the Akashic records. The Akashic records are like a library of all universal events, thoughts, words, emotions, and intents that have happened in the past, present, or future for all entities and life-forms in the universe. Theosophists believe the Akashic records are encoded in a nonphysical plane of existence they call the mental plane. There is currently no scientific evidence supporting the existence of the Akashic records, but consider that there was a time in our history when there was no scientific evidence for a lot of things people believed in without the scientific evidence to back it up that were later proven to be true. Imagine people's first reaction when it was finally proven that the earth is not flat or that the sun and stars do not revolve around the earth.

The pinpointing of when the universe began and how it all works together is mostly theories, based on our current understanding of time and space and of the laws of physics, mathematics, and all of science. In reality, the universe is a giant mystery with many more unknowns than knowns. This fact alone opens the door to the development of this new and untested hypothesis on how to establish meaningful communication with our ET brothers and sisters.

---

[1] Wikipedia, s.v. "Multiverse," last modified November 22, 2023, 09:59, https://en.wikipedia.org/wiki/Multiverse.

Much of our documented earth historical time line is considered to be verified by way of an analysis of physical artifacts in a process called radio carbon dating or just carbon dating. Based on this analysis, the dates of currently published historical time lines are considered to be mostly accurate, but they aren't always so. We must look more closely at what carbon dating is, understand its limitations, and determine if it is truly accurate and reliable in all cases.

Carbon dating has been generally accepted as the definitive method of analysis for closely estimating the age of any organic organism that once lived on our planet. We have put our blind faith in the words of the scientists who assure us that carbon dating is a reliable method of determining the age of almost everything that was once living. A closer look at the means and methods of carbon dating reveals that perhaps it is not quite as foolproof a process as we may have been led to believe.

There are two types of carbon found in organic material to the best of our current knowledge. One is carbon-12 (C-12), and the other is carbon-14 (C-14). To conduct a carbon dating analysis requires that the material to have been at one point organically alive to absorb carbon at a cellular level. Attempting to carbon-date rocks, ancient paintings on cave walls, or other inorganic materials is not possible.

The premise of carbon dating is that we know without question that all living things absorb both types of carbon, but once a living thing dies, it will stop absorbing. The C-12 is a very stable element and will not change form after being absorbed. C-14 is highly unstable and will immediately begin changing after absorption. Each nucleus will lose an electron, a process that is referred to as radioactive decay. This rate of decay has been proven constant and can be accurately measured in terms of a half-life.

Half-life is the amount of time it takes for an object to lose exactly half the amount of carbon stored within it. This half-life will continue at the same rate until the C-14 is gone. The half-life of C-14 is 5,730 years, which means that it will take that amount of time for the carbon to reduce from 100 grams to 50 grams, exactly half its original amount. In the same way, it will take another 5,730 years for C-14 to drop to 25 grams. By quantifying the amount of carbon stored in an object at the time of discovery and comparing it to the original amount of carbon *assumed* to have been stored at the time of the organism's death, scientists can estimate its age.

The issue is that the assumed amount of carbon existing in any living organism at the time of that organism's death is exactly that, an assumption. It is very difficult, if not impossible, for scientists to know precisely how much carbon would have originally been present in any organism at the time it died. One of the ways by which scientists attempt to compensate for this obstacle is to apply something called carbon equilibrium.

Carbon equilibrium is the point at which the rate of carbon production and the rate of carbon decay become equal. By measuring the rate of production and the rate of decay, which are both quantifiable, scientists assert that they are able to estimate that carbon in the atmosphere would go from zero to equilibrium in approximately thirty thousand to fifty thousand years. Since the universe is estimated to be millions of years old, it is assumed by scientists that this equilibrium has already been reached.

In the 1960s, the absorption rate of radioactive carbon was found to be significantly higher than the decay rate. This indicated that equilibrium had not in fact been achieved, which put into question the scientific assumptions about carbon dating. Scientists attempted to account for this by setting 1950 as a standard year for the ratio of C-12 to C-14 and by measuring subsequent findings against that.

The fact is that sometimes carbon dating findings will agree with other evolutionary methods of age estimation and that sometimes the findings will differ. When differences surface, scientists apply correction tables to reconcile results and eliminate discrepancies.

When standard carbon dating techniques do not concur with or contradict other estimates, carbon dating data is disregarded. This practice has been pointed out directly in the words of American neuroscience Professor Bruce Brew: "If a C-14 date supports our theories, we put it in the main text. If it does not entirely contradict them, we put it in a footnote. And if it is completely out of date, we just drop it." This means that carbon dating is not incontestable or reliable.[2] The point is that much of our true history and the age of ancient artifacts is probably not 100 percent accurate, for several reasons.

It can easily be argued that ETs have been coming and going from Earth throughout all of human history or, more likely, long before the time of humans. Seeking a message or code key about how to achieve two-way communication with other creatures of the universe beyond Earth seems complicated. Compared to understanding the complexities of the universe as a whole, seeking a simple message does not seem so daunting. To get there, we must rethink our means and methods and look back to the beginning of our learning curve on such matters. Something is clearly missing as we have apparently not yet achieved the goal of ET communication in modern times.

Many ancient cultures from all over the globe speak of portals to other worlds and gateways to star systems from where their descendants or perhaps their creators came from and reside. Recently declassified FBI files have openly stated that they believe that Earth has been visited by beings from other dimensional realms and planets. NASA has recently announced that they believe such portals do indeed appear to be hidden within Earth's magnetic field. The encouragement from NASA scientists makes it easier for both the scientific community and the religious community to consider that things like star gates, portals, and wormholes are real. These are some very intelligent and well-educated people making these scientifically risky claims without hard evidence.

How does the ancient and never-ending search for ET communication weave both spiritual faith and pure science? As for the position of science, NASA felt confident enough in the search to award $1.1 million to the Center for Theological Inquiry, which is an ecumenical research institute in New Jersey whose mission is to study the societal implications of astrobiology, which is a code word for intelligent ET life. Some hard-core scientists were very unhappy about this, as expected. The Freedom from Religion Foundation (FFR), which actively promotes the division between church and state, demanded that NASA revoke the grant and threatened to take legal action if NASA did not comply. While the FFR stated that their concern was the commingling of government and religious organizations, the FFR also made it clear that they considered the grant to be a waste of money.[3]

The day will inevitably come when humanity establishes communication with ETs. That accomplishment will raise numerous complex questions that will exceed the theological limitations of science. For example, when we ask, "What is life?," are we asking a scientific question or a theological question? Questions about life's origins and its direction into the future are complicated and must be studied equally across all

[2] "How Accurate Is Carbon Dating?" Labmate, May 20, 2014.

[3] "FFRF asks NASA to Withdraw Large Religious Grant," Freedom From Religion Foundation, June 9, 2016.

disciplines. This includes the establishment of ET communication and the discovery of life in the universe.[4] This is not just a fictional or esoteric fantasy. Many scientists now contend that the achievement of ET communication is more a question of when and not if. That is why the search continues at the great expense of time and money. We are living on the cusp of that achievement. These are exciting times in that regard.

There are numerous documented facts that justify such scientific confidence. One justification is the never-ending reports of legitimate UFO sightings. The primary justification, however, is arguably about the frequency at which scientists are identifying potentially life-supporting planets called exoplanets both within and outside our own solar system.

In the year 2000, astronomers reported that they had identified fifty exoplanets. By 2013, they had found almost eight hundred fifty exoplanets existing in more than eight hundred planetary systems. David Weintraub, associate professor of astronomy at Vanderbilt University and author of *Religions and Extraterrestrial Life*, says that this number may reach one million by the year 2045. He wrote, "We can quite reasonably expect that the number of known exoplanets will soon become, like the stars, almost uncountable." Of the exoplanets discovered thus far, more than twenty are Earth-sized exoplanets existing within the (Earth-like) habitable zone around their respective home stars. This includes the most recent exoplanet discovery in 2022, known as LP 890-9c or SPECULOOS-2c. Publishing their findings in the journal *Astronomy & Astrophysics*, researchers detected the first planet, LP 890-9b, around the star using NASA's ($287 million) Transiting Exoplanet Survey Satellite (TESS) project and followed it up using ground-based telescopes in the SPECULOOS consortium in Chile and Tenerife (in the Canary Islands). LP 890-9b orbits its star in the equivalent of 2.7 Earth-days.[5] Keep that $287 million expenditure in mind for when we get to the hypothesis and experiment proposed later in *Angel Communication Code*. That $287 million did not find ET life. It found a potential Earth-like planet, which is significant, but it fell far short of achieving ET communication.

The grand assumption beneath the exoplanet idea is that intelligent life will only be found on an Earth-like exoplanet. There is no reason to assume intelligent life must live on an Earth-like planet. It may very well live in a different environment with a different sort of atmosphere.

The SETI Institute (Search for Extraterrestrial Intelligence) is a not-for-profit research organization incorporated in 1984 whose mission is to explore, understand, and explain the origin and nature of life in the universe and to use this knowledge to inspire and guide present and future generations, sharing knowledge with the public, the press, and the government.

The SETI institute consists of three primary centers:

- the Carl Sagan Center, which is focused on the study of life in the universe,
- the Center for Education, which is focused on astronomy, astrobiology, and space science for students and educators, and
- the Center for Public Outreach, which produces the SETI's general science radio show and podcast, and *SETI Talks*, its weekly colloquium series.[6]

---

4 Brandon Ambrosino, "If We Made Contact with Aliens, How Would Religions React?," BBC, December 2016.

5 Jamie Carter, "A Newly Discovered Exoplanet Is Now Our Second-Best Target for Webb Telescope, Say Scientists," *Forbes*, September 13, 2022.

6 Wikipedia, s.v. "SETI Institute."

SETI's work on a day-to-day basis generally dwells within the boundaries of the traditional and known laws of science. The implications of SETI, however, extend far beyond biology and physics, reaching to the humanities and philosophy and even theology. As Carl Sagan pointed out in his book *The Cosmic Question*, "Space exploration leads directly to religious and philosophical questions." Carl Sagan was an American astronomer and science writer who popularized science through newspapers, magazines, television broadcasts, and books. He was a professor at Harvard University and the University of California, Berkeley, and a fellow at the Smithsonian Astrophysical Observatory and of the SETI Institute. He was controversial for his views on extraterrestrial intelligence, nuclear weapons, and religion. He was the narrator of the Emmy Award– and Peabody Award–winning PBS series *Cosmos*, which became the most watched series in public television history.[7]

A large part of this quest for ET communication is the need to take into account if our faith and beliefs will be synchronous with ET communication or if our faith and beliefs will be shaken irreversibly and fall to the ground. Either way, ET communication is going to happen someday soon. There comes a time when the questions about the existence of ETs end and we must study the ramifications of our inevitable communication with ETs and make that the priority. That time is now.

The theological pursuit of answers to questions about ET communication can be considered a new branch of science called exotheology or astrotheology. These terms were officially defined by Ted Peters, professor emeritus in theology at Pacific Lutheran Theological Seminary, to refer to "speculation on the theological significance of extraterrestrial life." The phrase itself dates back at least three hundred years to a 1714 publication entitled *Astro-theology or a Demonstration of the Being and Attributes of God from a Survey of the Heavens*. There are many questions and issues that establishing communication with ETs will bring to the surface when it happens. These include the question of our human uniqueness, which is an issue that has troubled both theologians and scientists for a very long time.

There are three principles guiding SETI, as Paul Davies explains in his book *Are We Alone?* First is the principle of nature's uniformity, which claims that the physical processes seen on earth can be found throughout the universe. This means that the same processes that produce life here on earth produce life everywhere in the universe. That seems logical.

Second is the principle of plenitude, which implies that everything that is possible will be realized eventually. That seems much less logical. It seems like a stretch to imply that anything is possible and, therefore, that everything that is possible will happen. For the purposes of SETI, the second principle claims that as long as there are no impediments to the forming of life, then life will form, or as Arthur Lovejoy, the American philosopher who coined the term, puts it, "no genuine possibility of being can remain unfulfilled." That is because, as Sagan claims, "the origin of life on suitable planets seems built into the chemistry of the universe." This second principle is a very complicated concept.

The third guiding principle is the mediocrity principle, which claims that there is nothing special about Earth's status or position in the universe. This principle provokes the greatest pushback from Christian religions. Christian teachings proclaim that the earth and human beings were purposefully created by God and occupy a privileged place above all other creatures in the universe.

In the modern era, there are vigorous protests from some religious groups about teaching scientific concepts such as evolution and the big bang in public schools. There are also the contrary but equally loud

---

[7] *Encyclopedia Britannica*, s.v. "Carl Sagan."

proclamations from some people of science with personal antireligious philosophies. It would seem that science and religion will be at odds with each other until the view of either one side or the other is proven beyond contestation. News outlets are quick to publish reports over which scientists and religious leaders launch attacks at one another. But just how representative are such conflicts? In reality, the media hype given to such episodes camouflages the far more numerous cases where science and religion harmoniously and synergistically coexist.

The more common dynamic is that people of many different faiths and levels of scientific expertise see no contradiction at all between science and religion. Many simply acknowledge that the two institutions deal with different realms of human reality and the human experience.

Science investigates the natural world, whereas faith deals with the spiritual and supernatural. Therefore, the two can be complementary. Many faith-based organizations have issued statements declaring that there need not be any conflict between religious faith and the scientific perspective of evolution. Contrary to the stereotype, one certainly does not have to be an atheist in order to be a scientist. Science does not prevent a scientist from having faith in a deity. A 2005 survey of scientists at top research universities found that more than 48 percent of them have a religious affiliation and more than 75 percent believe that religions convey important truths about the universe. Some scientists, such as Francis Collins, former director of the National Human Genome Research Institute, and George Coyne, astronomer and priest, have been outspoken about the satisfaction they find in viewing the world through both a scientific lens and one of personal faith. Obviously, there will always be certain subjects where theological and scientific disagreements will arise. For example, when strict Christian tenets contend that the universe was created in six days as some literal interpretations of the Bible might require, faith and science can find themselves in conflict, considering the overwhelming magnitude and lack of hard evidence of such a creation in so short a time. The concept of time has baffled people for millennia, and still to this day time is not fully understood or defined.

Albert Einstein made a claim that time is not regular at all. He said that time is variable in its nature and depends on relative speed. For example, imagine you are in a bus that is traveling at just a fraction below the speed of light. According to Einstein's special relativity theory, the rate at which time travels on the bus depends on whether the bus is being observed while sitting outside that bus or on the bus.

If you are sitting inside the bus, you might feel that time is passing quite normally; however, if you were sitting outside the same bus, your perception of time would assume a different quality. Now looking inside the bus from the outside, you would perceive time differently. This is because, according to Einstein, the faster the object moves, the slower the time will pass for someone observing from the sidewalk. Therefore, time varies, and it is just a matter of speed.

Second Peter 3:8 in the New Testament says: "But, beloved, be not ignorant of this one thing, that one day is with the Lord as a thousand years, and a thousand years as one day." How could Peter, who was less scientific in his approach, say something so scientific that took almost two thousand years for someone such as Einstein to discover? Was it a simple prophetic vision that made Peter say that?

For most of us who do not possess the scientific genius of Einstein, we search for God because we assume God is outside of the world. Therefore, it is admissible that for God, things are moving slower because observation is taking place from the outside. But God is inside the world for most of us, after every thousand years, and first day completed.

Another important verse in this context is in the Old Testament, Psalms 90:4. It reads, "For a thousand years in thy [God's] sight are but as yesterday when it is past, and as a watch in the night."

This verse supports the claim made by the apostle Peter and Albert Einstein about time, because David is writing the psalm almost one thousand years before Peter's announcement about time. It would be reasonable to suggest that when Einstein was grappling with the concept of time, he well knew of the Jewish writers' statements about time, including those of King David and Saint Peter Simon. Einstein was working on a concept that had been presented twice before him. It would not be incorrect to say that he did not conceive the concept. What he did do was to take the existing concept and apply scientific law.

It is quite curious that King David also looked to history for a source to understand time in the same way that King David's writings were the source and reference point for Saint Peter Simon. Further, perhaps both their writings were the reference points for Einstein.

Three people said the same thing about time at different intervals in time for the civilizations of their respective times. It has only to say time is temporary with a beginning and an end. It all means that the current time of today is more or less the sixth day described in the book of Genesis,[8] the time when God said he was done.

Albert Einstein's religious views have been widely studied and are often misunderstood. Albert Einstein himself stated, "I'm not an atheist, and I don't think I can call myself a pantheist .... I believe in Spinoza's God who reveals himself in the orderly harmony of what exists, not in a God who concerns himself with fates and actions of human beings." Einstein believed the problem of God was the "most difficult in the world" and a question that could not be answered "simply with yes or no." He conceded, "The problem involved is too vast for our limited minds."

Einstein explained his views on the relationship between science, philosophy, and religion in his lectures of 1939 and 1941: "Science can only be created by those who are thoroughly imbued with the aspiration towards truth and understanding. This source of feeling, however, springs from the sphere of religion" because "knowledge of what is does not open the door directly to ... what should be the goal of our human aspirations." All the aspirations "exist in a healthy society as powerful traditions," which "come into being not through demonstration but through revelation, through the medium of powerful personalities. One must not attempt to justify them, but rather to sense their nature simply and clearly. The highest principles for our aspirations and judgments are given to us in the Jewish–Christian religious tradition."[9] These are powerful and very relevant insights to mull over, particularly when we consider the moral failures seen and even accepted in the world today. The next move in God's playbook is the Revelation—and the book of Revelation says much about interactions with beings not of this earth.

It is critical for humanity to understand that behind the scenes and out of the spotlight, many cases exist where religious and scientific perspectives present no conflict at all. Thousands of scientists carry out their research while maintaining personal spiritual beliefs. Many people view the natural world through

---

8 Robin Kumar, "How 1,000 Years Equals to 1 Day for God?," Reflection, May 12, 2018.

9 Wikipedia, s.v. "Religious and Philosophical Views of Albert Einstein."

an evidence-based scientific lens and also through a spiritual lens. It is a difficult but achievable balance. Accepting a scientific worldview need not require giving up religious faith, or vice versa.[10] As is the case with most arguments, the truth, more often than not, lies somewhere in the middle. What and where is the middle? Messages from angels may very well be that middle ground. They deliver divine messages, and at the same time they may be delivering a less conspicuous message about ET communication via their structural hierarchy and actions. Let us not forget that all the angel messages we look back upon throughout Christian history and as written in the Bible are written by human authors of those ancient texts. The Word of God is documented by human interpretation and their recording of extraordinary events.

Cosmology could also be viewed as a middle ground field of study linking religion and science. Cosmology is the branch of astronomy that seeks the origin and evolution of the universe from the theoretical big bang through today—and on into predictions for the future with respect to the behavior of cosmos. As long as we have been trying to understand the universe, we have been developing and testing cosmological theories. The inclusion of a deity or higher power frequently is woven into these cosmological theories. In most monotheistic religions, the subject god is the one and only creator and controller of the universe. In the past hundred years, however, a different sort of cosmology has emerged. Scientific cosmology has traditionally incorporated mathematics and physics, but now it openly incorporates theology into its considerations and theories.

Despite the fact that many of the Christian faith have engaged theologically with cosmology, it has not often been the case that the defining feature of Christianity has been an explicit part of these interactions.[11] In our quest to find an ET communication code involving angels in all this complicated cosmology and theology, we need to go back to basics and think again. We need to understand how we got to where we are today with respect to our understanding of science, faith (angels), and the universe (ETs).

Going back in time as far as we are able, we know that humans have always felt a connection to the stars. Many ancient cultures believe we came from the stars. Over time, the repetitive and consistent movements of stars were studied to develop a different level of understanding of the night sky, thereby making astrology and astronomy the oldest sciences. The human story of science and faith begins with that and leads up to the story of the star that guided the three Magi (Wise Men) to the manger where Jesus, the future King of angels, was lying after having been born. It is known as the star of Bethlehem. This is one of those instances where science and faith do not always agree. Scientists have been trying to find a non-Christian explanation for this event for a very long time. We may never know for sure what the star of Bethlehem was or if it even really happened, but the story exists in the Bible for a reason. The question comes up year after year. We may never know if the star of Bethlehem and the story of the Magi was a scientifically explainable event or if it is one of those self-serving manipulations of history told to advance the cause of Christianity.

The earliest humans looked up into the night sky and made up their stories about what all that stuff up there was about; these stories were based on nothing more than what their imaginations could conjure. It was not until humans began to record the patterns of celestial motions that we began to formulate hypotheses leading to explanations of the movements of stars in those times. It is believed that the first known recording of the motions of the stars was approximately seven thousand years ago at a site called

[10] "Science and Religion: Reconcilable Differences, *Understanding Science*, Berkeley.edu.

[11] *Stanford Encyclopedia of Philosophy*, ed. Edward N. Zalta (Stanford, CA: Stanford University, Winter 2021), s.v. "Cosmology and Theology," Halvorson, Hans, and Helge Kragh.

Nabta Playa. Based on those early recordings and subsequent celestial observations, the indication is that for thousands of years, ancient societies all over the globe proceeded to construct gigantic stone circles and monoliths, aligning them with the sun, moon, planets, and stars. The true purpose of these constructions is unknown. It has been hypothesized that these structures were built to predict the coming of seasons for rain, planting, and harvesting and the impending migration of animal herds. This begs the question of why. No other creature on earth, including creatures of the oceans, needs such devices to know the rainy season is coming or that it is time to migrate. They just know. We also speculate that these structures served as ceremonial sites for celebration and ritual sacrifice. That makes some sense. What makes even more sense is that the humans who built them were searching for a way to connect with the stars. What is relevant to *Angel Communication Code* is that we believe ancient peoples did not have the means or the methods to construct these sites on their own. They most certainly were communicating with ETs that did what was necessary to get these structures built and positioned in a astrologically correct way for a purpose. That ultimate purpose remains a mystery. In the overwhelming majority of instances, the boulders that make up these giant structures are cut and placed with laser-like precision. And not only that, but also they weigh hundreds of tons. These structures were obviously constructed with a superior knowledge and a technology that ancient humans simply did not possess.

Europe holds approximately thirty-five thousand megalithic structures, most of which are astronomically aligned stone circles. It is believed that the majority of these structures were built between sixty-five hundred and forty-five hundred years ago, along the Atlantic and Mediterranean coasts. The most famous of these sites is probably Stonehenge, a site in England that is thought to be approximately five thousand years old. It is currently believed that Stonehenge could actually be one of the youngest of these sorts of structures ever built in Europe.

The Nabta Playa site, cited previously, is located approximately seven hundred miles south of the Great Pyramid of Giza in Egypt. J. McKim Malville, University of Colorado professor emeritus and one of the world's top archaeoastronomy experts, once said of Nabta Playa, "Here is human beings' first attempt to make some serious connection with the heavens."

Humans were evolving from curious stargazers into scientists. Astronomy, an accepted scientific discipline, is the study of phenomena and objects beyond Earth's atmosphere. Astronomy emerged in China, India, Egypt, Europe, Mesoamerica, and the Middle East. Astrology, on the other hand, is the study of planetary positions to predict the future of our physical environment and also what the future may hold for a person. Today, astrology is considered a pseudoscience at best. This, however, was not always the case. Before the seventeenth century, astrology and astronomy were considered a single discipline. Astrology was viewed as a form of applied astronomy, and its predictions were believed to be very real and very important.

The disciplines of astronomy and astrology have a common origin in ancient Babylonia between approximately 3000 BC and 2000 BC. The Babylonians needed a detailed celestial calendar to predict the rising waters of the Tigris and Euphrates Rivers (in present-day Iraq) for their floodplain agrarian agriculture, and based on that need, they developed into masters of astronomical observation in those days. Their pursuit of astrology, however, was rooted in religion. Although the stars follow a regular course from east to west as the night progresses, the planets do not. They appear to move backward and then forward when their positions are tracked and recorded. This erratic motion of the wandering planets that were known in the time of the Babylonians (i.e., Mercury, Venus, Mars, Jupiter, and Saturn) caused the Babylonians to believe that the planets were gods possessed of their own powers. In approximately 500 BC, Babylonian priests identified the constellations that mark the zodiacal band still in use today. For the Babylonians, the zodiac not only was a reference marker for charting the motions of the sun, the moon, and planets but also was used for charting the position of the planet-gods in the zodiac's constellations, which were used for astrological predictions.

The casting of zodiac horoscopes first occurred, as far as we know now, in Mesopotamia during the Persian occupation in 450 BC. The first-known cuneiform horoscope was cast in 410 BC. The cuneiform horoscopes are based on applying the situation of the heavens when a person is born to the individual's future life etc. These horoscopes predicted a child's future and personality, and the length of his or her life, based on planetary positions when the child was born. The horoscopes took into account the idea of twelve astrological houses. Each house represented some quality or aspect of human life. For instance, the first house predicted personality, and the second predicted finances. The ascendant, the zodiacal sign rising over the eastern horizon at the time of birth, was placed in the first house. The ascendant subsequently determined the locations of other celestial bodies in the other houses, which together composed the individual's horoscope.[12]

Ancient humans recorded astrological observations and noted changes, then linked those changes to the behavior of the observable world. This is the very essence of the contemporary scientific method. Civilizations around the world in those days depended upon those who could interpret the motions of the night sky. Civilizations of ancient times genuinely needed astrologer-astronomers.

The history of Western astronomy is rooted in ancient Mesopotamia. Astronomy appeared alongside the development of agriculture in a place called the Fertile Crescent, which is believed to be the birthplace of many things, including both farming and writing.

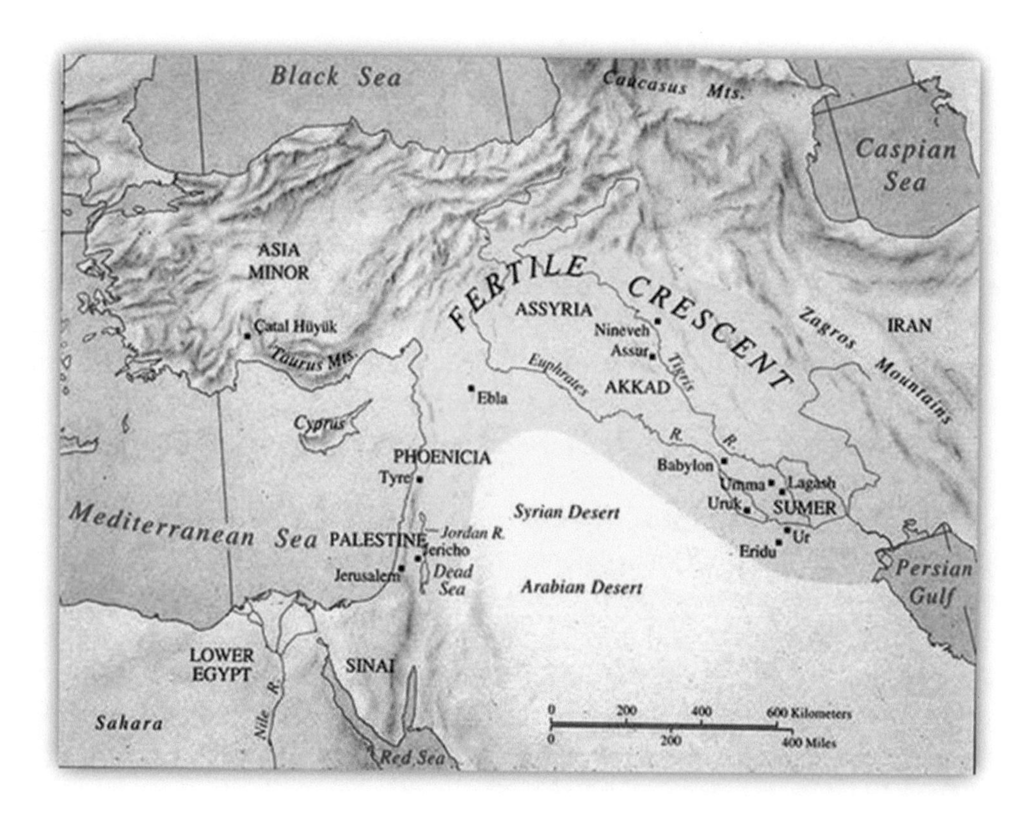

---

12 Encyclopedia.com. Steven V. Roberts, "Astronomy Emerges from Astrology," Encyclopedia.com, May 1988.

As ancient civilizations rapidly grew in Sumer, Assyria, and Babylon, so did the study of the night sky. Although it is true that modern Europeans may have adopted the constellations identified by the Greeks, knowledge of those constellations was already ancient in the days of Aristotle. This means that we can trace the origins of today's constellations all the way back to the time of the Babylonians.

The Babylonian people had a unique tradition of developing star maps. They kept track of two separate sets of constellations for two distinctly different purposes. One set was used to track farming dates and mark the time for ancient celebrations. The other was devoted to recognizing the gods. This god-recognition map made its way to the Greeks, forming the foundations of our modern twelve constellations of the zodiac. All this is important to the premise of *Angel Communication Code* because the number 12 has a spiritual, biblical significance, including as a reference to angels.

The number 12 is cited one hundred eighty-seven times in the Bible. The Old Testament book of Genesis tells us that Jacob had twelve sons, who later formed the twelve tribes of Israel (Genesis 49:1–28). Also, the breastplate of decision, otherwise known as the high priest's breastplate, was embedded with twelve stones representing the twelve tribes of Israel (Exodus 28:15–21). Additionally, Shem, one of Noah's sons saved from the Great Flood by the ark, is among the twelve patriarchs cited in the Old Testament.

In the Bible's book of Numbers, upon their escape from Egypt, the Israelites wandered in the desert. They sent out twelve spies to cast about the Promised Land. It's also worth noting that Elijah built an altar consisting of twelve stones, with each representing one of the twelve tribes of Israel (1 Kings 18:31).

The first words Jesus spoke about God in public were said to have been spoken when He was twelve years old (Luke 2:42–51).

In the holy Gospels, we are told about the twelve apostles and the authority they were given by Jesus as their leader. They were known as the twelve disciples in the beginning but later became known as the twelve apostles. In addition, when Jesus fed the five thousand, the disciples gathered twelve baskets of leftover food (Matthew 14:20).

In Revelation 21:9–21, the foundation of the holy city in heaven, New Jerusalem, is ornamented with twelve beautiful stones, and there are twelve gates surrounding it with twelve angels guarding these gates. Each gate is named after one of the twelve tribes of Israel, which are named after the twelve sons of Jacob. In the next chapter, John also sees the tree of life, which bears twelve different kinds of fruit that grow every month (twelve times per year) (Revelation 22:2).

Last but not least, the number 12 is also associated with an anointed service. In the Bible, twelve men were anointed by a prophet of God to carry out a unique task. These were Aaron along with his four sons (anointed into the priesthood in Exodus 29:7–9) and Saul; David; David's son Absalom; King Jehu, Kings Joash and Jehoahaz of Judah; and King Solomon.

The point is that 12 is another number that was historically put in play for a reason. It is not coincidental or random. There is a code or pattern out there for us to decipher that is linked to the heavens, the stars, and the universe. Angels and scripture play a key role in the message's position.

Humans have been in search of ET communication since the beginning of our history. There is evidence to suggest that ET communication occurred many times in antiquity; however, the lines of communication were largely cut off sometime around the days of Noah. Many ancient cultures believe they are descended from star people; others believe they were created by a higher power not of this earth. How to reconnect with that ancestry, by whatever means, is the ancient issue that remains to this day. The search for answers began by trying to understand not only the stars, but also ourselves.

The Babylonians did not merely draw pictures of the night sky. They also chiseled them into rock. By thirty-two hundred years ago, they had carved the first (known) catalog of stars into stone tablets. The names given to some of those stars may have even older origins, apparently coming from the Sumerian people. This implies that formal knowledge of the stars stretches back to before recorded history. These developments were not unique to the West. Similar histories played out on different time lines in cultures all over the world. These facts are the justification that many historians use when considering astronomy, inclusive of astrology, to be the oldest science.[13]

Examining the history of astrology opens the door to discussion about mystics. Just saying the words *mystic* and *astrology* in modern times tends to conjure up a stereotype, causing many people to become immediately skeptical about what peculiar discussions might follow.

It was actually a mystic who defined the accepted Christian angel hierarchy as we know it and apply to Christianity today. It did not come from the Bible. This will be discussed in greater detail later in *Angel Communication Code*. Mysticism has been called "the science of the love of God" and "the life that aims at union with God." Mystics are present in every religious tradition, sometimes as central participants, but often on the fringes of accepted practice because they describe a less traditional experience of divinity.

There is no identifiable mystic type. Mystics may be women or men, educated or uneducated, from wealthy or impoverished backgrounds. Mystical experiences can be visual, auditory, or of the mind only. Mystical experiences can occur spontaneously and unexpectedly, at any time and any place. Many religions encourage specific practices and methods of prayer that instigate mystical experiences. All traditions seem to agree that mysticism is an unusual gift that is not fully understood and is not fully under the control of the gifted. During some historical periods, mysticism was much more prevalent and more authoritative, and mystics were much more respected and revered by their communities.[14]

A mystic can be described as a person who pursues a truth or understanding beyond those truths normally associated with the human experience. Mystics may or may not be initiated into any number of spiritual or religious mysteries. They may or may not have achieved the insight they seek. What links all such people together is the belief in, and pursuit of, a transcendent truth that goes beyond rational, rock-solid understanding. In today's popular understanding, a mystic is generally a person who deals in abstract practices or who studies magic or the occult. Mysticism has a surprising number of facets, and trying to define or understand it beyond a pursuit of transcendent truth can be quite a complex task.

Nearly all religious traditions have their own strains of mysticism and astrology. In many monotheistic and some polytheistic faiths, a mystic person is usually focused on finding a direct connection to God Himself, largely through meditation or prayer or both. In Christianity, mystics often refer to this state as "union" or "oneness with God." In Islam, this state is called Irfan, which literally translates as "knowing." In Jainism, this union state is called moksha, referring to an ascendance to a spiritual state where all reality

---

[13] Eric Betz, "Why Astronomy Is Considered the Oldest Science," Astronomy.com, October 6, 2020.

[14] Elizabeth Alvilda Petroff, "The Mystics," *Christian History* 30 (1991).

is considered an illusion. It is all connected to communication with beings not of this earth, which is what *Angel Communication Code* is focused on.

Within many mainstream religions, there are varying degrees or levels of mysticism. Some of these religious groups are considered completely unholy by mainstream religious faiths. Many modern mystic practices are heavily influenced by ancient Greek traditions such as the Eleusinian Mysteries, which date back to the fifteenth century BC. The Eleusinian Mysteries focused on a myth cycle involving Demeter and Persephone, invoking the concept of death and of the resurrection that can come by triumphing over death. They remained in place for nearly two thousand years. Over that time, they laid much of the foundation for the mystical practices other faiths would adopt.

From the seventeenth century on, various fraternal organizations that incorporate mystic elements began to gain popularity throughout Europe. The Rosicrucian Order and the Freemasons are perhaps the best known of these groups, and they continue to be popular among a widespread demographic to this day. The following is taken from the Rosicrucian Order's website:

> The Rosicrucians are a community of philosophers who study and practice the Natural Laws governing the Universe. Our mission is to provide seekers with the spiritual wisdom necessary to experience their connectedness with the miraculous world around us and to develop Mastery of Life. Mysticism teaches cosmic laws and principles by which we are brought into closer consciousness of our divine power. The mystical experience of union with the One imposes upon the mystic a moral obligation—to use this knowledge for the welfare of others. Through our teachings you will gain specific knowledge of metaphysics, mysticism, philosophy, psychology, parapsychology, and science not taught by conventional educational systems or traditional religions.

Some of the more prominent members of the Rosicrucian Order are (or were) George Washington, Abraham Lincoln, George W. Bush, Barack Obama, Benjamin Franklin, Walt Disney, Isaac Newton, Napoleon Bonaparte, Francis Bacon, René Descartes, Saint Thomas Aquinas, and Leonardo Da Vinci, to name just a few. The Rosicrucians claim to currently have more than ninety-five thousand members across the globe, and their membership is growing.

Early in the nineteenth century, there was a resurgence of mysticism in the West. These factions often incorporate occult elements such as communication with spirits.[15] What is not talked about much within the Christian community, including Christian teachings, is that mysticism is a Bible-based religion. Every committed Christian who has genuinely been, or seeks to be, touched by God is a mystic by definition. There are absolutely no exceptions. Mysticism is central to the messages of the Bible and angels. All Christians either are mystics or follow a faith that is entirely different from biblical Christianity. The cornerstone of Christianity and all its forty-five thousand denominations across the globe is "the mystery of faith." The same can be said for any religious faith with a supreme deity. Many religious believers have problems with mysticism at face value, mostly because they do not understand what mysticism is or its priority, purpose, or potential.

Mystical experiences are frequently referenced in the Bible. Enoch walked with God. Moses had his burning bush. Abraham entertained angels. Gideon spoke with God. The Virgin Mary, mother of Jesus, spoke with an

[15] Brendan McGuigan, "What Is a Mystic?," PublicPeople.org, August 24, 2022.

angel, as did many others. Christian mysticism is rooted in Trinitarian life, that is, the relationship between Father, Son, and Holy Spirit, in which every Christian is required to believe. The Trinity is, as stated by one unknown Puritan writer, "the life of God in the soul of the Christian man." Christian mysticism is delivered by God's established means and methods, which include angels. People of faith want to know intimacy with their god. People of faith are mystics by their very desire to exist within their god's eternal presence.[16] It was mystics that were the interpreters of astrological science and predictions in ancient times. That was their purpose and focus. They are also deeply rooted in Christianity and are connected to the defining of the Christian angel hierarchy.

The growing rift between astrological and astronomical beliefs came to a head in the sixteenth century, beginning with the work of German astronomer and astrologer Johannes Kepler (1571–1630). His scientific discoveries combined with cultural factors led to the full-blown separation of astrology from astronomy as independent scientific disciplines by 1700. That was only 323 years ago!

Documented history is authenticated by the verification of critical sources of information, the selection of specific pieces of information from source materials, and the assembly of those particulars into an accounting of historical events. The issue is that as new information is discovered postpublication of an original document, recorded history is often proven to be incorrect based on this new finding. In other cases, recorded history is found to incorrect or incomplete on account of the author's bias. Consider two fundamental necessities with respect to the recording of religious history and the messages contained therein:

1. Historical religious writings redacted relevant events that did not support their prevailing religious system or message.
2. Religions needed self-justifying histories, meaning that significant historical events had to have a place in the scheme of things that led to the ultimate enlightenment that these religions claimed as fact.

Religious teachings needed a philosophy for the respective religion to defend itself against, or to gain favor with, those people to whom religious philosophy was important. Religions legitimized the use of philosophy by giving it a history, which limited the recording of events to those that pointed people in the right direction and enabled some understanding of the beliefs and actions of the people who lived before their time.

All religions filtered and documented their histories in a way to focus on continuity and the guiding of the people toward the respective religion's course to enlightenment according to their specific religious beliefs. The premise of *Angel Communication Code* recognizes this dynamic in claiming that angel messages had two purposes. One was about the Christian faith, and the other was about finding the path to ET communication. It can be argued that this dynamic is the root cause of the rift between religion and science. We must recognize and understand this dynamic and not allow the self-servingly documented religious history to lead us into thinking that there is a long history of confrontation between religion and science. The contemporary conflict between faith and science is about the belief in holy creation after Charles Darwin's *On the Origin of Species*. Darwin's work provided a foundation for scientists to promote their conjectures over the creationists' views.

---

[16] Donald Richmond, "Every Christian Is a Mystic," August 31, 2015.

We possess only carefully selected pieces of evidence about the true lives of the ancient people of faith and science. The written history that has survived, has survived precisely because historians have chosen not to question that history, or have accepted the history in order to sell us the story, be it fully complete or not. The survival and curation of historical information tells us much about the selectors and keepers of that information and perhaps not enough about the original subjects themselves.

Despite the degradation of the scientific validly and status of astrology over the years, astrology has proven to be an incredibly resilient discipline. At the end of the twentieth century, when a classic study of ancient astrology was written by Auguste Bouché-Leclercq, contemporary astrology looked as if it had permanently disappeared. He seemed to be writing about the history of an ancient and extinct superstition. Soon after that publication, astrology entered somewhat of a renaissance period. The primary mechanism of this resurrection can be credited as being the newspaper horoscope. Modern horoscope predictions, however, are about as far from ancient astrology's study and purpose as horoscope predictions can get. A modern horoscope predictor is more like a psychological counselor, concerned for the emotional health of his or her client. The most prolific consultant astrologers today set themselves up as self-qualified counselors who combine knowledge of astrology with a background in psychotherapy or psychology. Ancient believers in astrology did not want counseling. They wanted predictions and help with making specific life-altering decisions.[17]

The ancients looked to the stars to find answers with the best scientific methods they had at their disposal, which was the observation and recording of the patterns of movement of the stars, which move in measurably predictable cycles over time. Modern horoscopes have degraded into a completely different thing with a different purpose. For the purposes of *Angel Communication Code* regarding the human search for ET communication, however, ancient original astrology has some significance on several levels, which will be later revealed.

The entire purpose of this discussion on astrology with respect to the subject of *Angel Communication Code* is to demonstrate that in ancient times, astrology and the mystics were much more dialed in to the importance of messages from the heavens than we are today. The ancients have proven to be reasonably on-target as the centuries have passed and we can now look back. We can look back and assess what was predicted, how it was predicted, and what the results of those predictions were. To figure all this out, we must go back to basics to understand the ancient astrological facets of mysticism and their relevance to communication with ETs.

Astrological ages are based on the precession of the equinoxes and the counterclockwise wobble of the earth's axis through the constellations, which lie in a circle around the earth along the ecliptic, which is the latitude of the equator, which is at 0 degrees. The 23.5-degree wobble takes approximately 25,920 years to complete, and in one complete cycle, there are twelve ages of 2,160 years each. That marker currently points to the constellation Pisces. Therefore, we are currently in the Age of Pisces, the twelfth and last sign of the zodiac cycle.[18] Recall our previous discussion about one day for God being a thousand years to us, thus placing us in the sixth day. This is similar to our being in the last sign of the zodiac today. This means that the previous, the current, and the next zodiac ages were, are, and will be Aries (1), Pisces (12), and Aquarius (11).

---

17 Tamsyn Bronson, *Ancient Astrology* (London: Routledge, 1994).

18 Robert Fitzgerald, "Astrological Ages as an Accurate and Effective Model of History," *Astrological Journal* (2009).

The twelve signs of the zodiac "eras" that progress in a clockwise, chronological order are as follows:

1. Aries (March 21–April 19)
2. Taurus (April 20–May 20)
3. Gemini (May 21–June 20)
4. Cancer (June 21–July 22)
5. Leo (July 23–August 22)
6. Virgo (August 23–September 22)
7. Libra (September 23–October 22)
8. Scorpio (October 23–November 21)
9. Sagittarius (November 22–December 21)
10. Capricorn (December 22–January 19)
11. Aquarius (January 20–February 18)
12. Pisces (February 19–March 20)

Following is a typical image of the zodiac wheel:

Wherever you stand on the past, present, or future significance of astrology and the ages of the zodiac, the validity of it all, consider the following about the astrological predictions for the previous, current, and next ages of astrology, the first being science. The three ages most relevant to us today are the immediate past, the present, and the immediate future (Aries, Pisces, and Aquarius). *Angel Communication Code* is not intended to be a history lesson skewed toward the legitimization of astrological predictions to justify the findings cited herein. There is, however, a distinct correlation between what was predicted and what actually happened in each historical age.

It is important to understand that there is no formal consensus among astrologists with respect to the exact dates of the ages. One of the more prominent theories and opinions on this matter is that the first age of the last full cycle, Aries, ends with the birth of Jesus Christ. Therefore, the Age of Aries began 2,160 years prior.

1. The Age of Aries: 2160 BC to 0 BC/AD

Aries is the ram and is said to rule warfare, the overcoming of challenges, competition, anger, aggression, and independence. If you research the history, you will find that the previous Age of Aries was clearly influenced by those things that were predicted. The Age of Aries, the ram that rules warfare, ends with the birth of Jesus the Christ, the Prince of Peace, the King of angels.

2. The Age of Pisces: 0 BC–AD 2160

We are currently approaching the end of the Age of Pisces. Pisces is the twelfth and final sign of the zodiac wheel or first age (counterclockwise from Aries). The sign of Pisces is two fishes. Pisces rules religion and spirituality, transcendence, prophecies, and prophets, among other things. History shows that the previous Age of Pisces was the age of great world religions, which are Judaism, Christianity, Islam, Hinduism, and Buddhism. Judaism transformed itself into a world religion after the prophetic revolution in the eighth and seventh centuries BC. All the major world religions were developed during this age, as predicted.

The heart of these religions, and especially of Christianity, has been the message of prophecy. This message of end times is fitting in the Age of Pisces, the last sign in the zodiac. This is the story told in Revelation, the last book of the Bible's New Testament. It is the apocalypse at Armageddon. There is a spiritual battle between God and Satan, or between Ahura Mazda and Ahriman, or between enlightenment and release from reincarnation and the darkness of imprisonment on the wheel of life. At the end of this cosmic battle, there will come a new cycle that heralds in the golden age of promise under the sign of Aquarius.

3. The Age of Aquarius (AD 2160–4320)

The coming Age of Aquarius is predicted to begin approximately 137 years from 2023. Aquarius is the immediate future of humankind. Aquarius is one of the signs of water, which is the most essential ingredient of life as we know it. The Age of Aquarius highlights the qualities of genius, science, knowledge, humanitarianism, brotherhood, and infinite possibilities. This is consistent with the establishment of two-way communication with the community of the universe. It is our inevitable and impending future.

We cannot look back to see what was going on in the world during the previous Age of Aquarius nearly twenty-two thousand years ago, as recoded history does not go back that far.

The point of all this discussion of the zodiac is to demonstrate that the historic record demonstrates that there is an uncanny truth hidden within the ancient astrological predictions that have played out since the emergence of astrology as the original science. It cannot be dismissed without being given due consideration. We must factor in what it might all mean with respect to our connection to the universe and our impending communication with the universal extraterrestrial community of intelligent life. We are right on the cusp of making this happen. We live in an exciting and very important time in history.

After all these years, wherever you have been sitting on the theological or scientific fence, you, as do we all, still ponder several fundamental concepts that we have failed to prove or disprove beyond any doubt. All these discussions as we zigzag back and forth between science, faith, angelic messages, visions, and codes will be linked in the end. *Angel Communication Code* is building the foundation that will justify the execution of an experiment via the identification of coded clues to a message that may very well close the communication gap between humankind and extraterrestrials.

CHAPTER

2

Communication Codes and Messages

Creating and deciphering codes and messages is hardly a new thing. Since humankind first began roaming the earth, we have sought messages, codes, clues, and patterns to describe the natural world as a matter of importance to our immediate survival. We are also searching for an understanding of what makes our earth and the universe function. In addition, we seek an understanding of ourselves and our purpose and place in the universe. It is human nature to seek answers to the unknown. We don't like the unknown because we can't control the unknown, and therefore it frightens us. No creature anywhere in the universe enjoys feeling afraid.

Everything about the structure of all life on earth as we currently understand it is based on a biological code. The DNA (deoxyribonucleic acid) code is the blueprint of all life here on earth. We assume this is also true for all life in the universe. DNA was discovered by a Swiss physician named Friedrich Miescher in 1869. He discovered it but did not know exactly what it was or how it worked.[1] In 1953 James Watson and Francis Crick became the first scientists to formulate an accurate description of the DNA molecule's complex double-helical structure. It was later discovered that DNA has a code language that can be deciphered using only four letters, which make up "codons." Codons are words, each three letters long. Deciphering the language of the genetic code was the work of Marshall Nirenberg and his associates at the National Institutes of Health in the 1960s. They found that the way for interpreting the entire human genome was to solve a simple code.[2] The DNA molecule and its structure are extremely complex; however, DNA's code is modeled using only four letters of the alphabet. The biological code describing all life on earth is ultimately described by a simple code. The trick was figuring out the key to the code, but there is even more to it than just that.

1 Cynthia McKanzie, "Mystery of Our Coded DNA—Who Was the "Programmer?," Message to Eagle, May 10, 2017.

2 American Chemical Society, *Deciphering the Genetic Code* (Bethesda, MD: National Institutes of Health, November 12, 2009).

Following is a typical diagram of what a DNA molecule looks like:

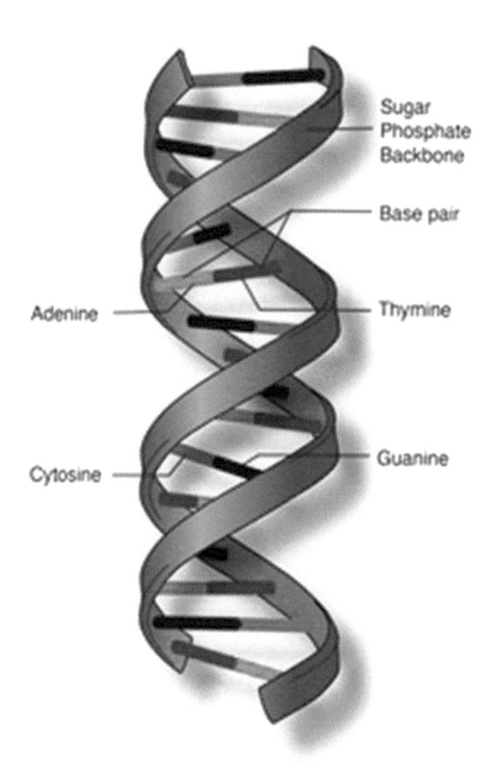

The four nitrogenous base acids (adenine, thymine, cytosine, and guanine) appear in a specific repeating pattern.

Every ten acids in sequence, there is a bridge, then every five acids, there is another bridge, then every six acids there is another bridge, and finally there is another five acids. Then this pattern repeats. Every DNA molecule follows this 10–5–6–5 pattern.[3]

In the ancient Hebrew language, these numbers correspond to these letters: 10 = Y, 5 = H, 6 = W. Therefore, the 10–5–6–5 pattern is equivalent to YHWH.

Yahweh is name of the God of the Israelites. The biblical pronunciation is "YHWH," the Hebrew name revealed to Moses in the book of Exodus. The name YHWH, consisting of the sequence of four consonants, yod, heh, waw, and heh, is referred to as the Tetragrammaton.[4]

The name Yahweh or YHWH is first mentioned in the Bible in Exodus 2:4. It is used next in Exodus 3, where Moses records the account of receiving his commission from God. In verse 15, God refers to Himself as "YHWH Elohim," translated "Lord God" in most English-language Bibles. In the Authorized Version, wherever we find the name "God" or "Lord" printed in small capitals, the original is "YHWH." The New Living Translation Version (NLV) Bible uses the term "Lord God." In the original Hebrew text, the name used for Lord in the verse is Yahweh. In this first usage of the word "Yahweh," it establishes that He is the God and the Creator of the universe. It states that He made the heavens and the earth. The name Yahweh, which is

---

3 "Stunning! The Creator's Name Is Found in Your DNA!," Bridge Connector Ministries, April 12, 2021.

4 *Encyclopedia Britannica*, s.v. "Yahweh."

usually spelled as "Jehovah" in English translations of the Hebrew Bible, appears 6,519 times. In the original Hebrew Bible, the name YHWH occurs 6,800 times.[5]

Just like DNA three-letter codons, the name YHWH is made up of a sequence of four syllables of three letters each: yod, heh, waw, and heh. As Judaism grew into a more universal, rather than a small local, religion, the more common Hebrew noun *elohim* (plural in form but understood in the singular), meaning "God," tended to replace YHWH. This change started sometime during the Babylonian exile.[6]

The point here is that the original name of God, YHWH, is coded into the patterns of every DNA strand. How did that happen? God gave Moses the term YHWH and God created the DNA code, whose pattern returns the name of God, YHWH. This is an obvious, powerful, and meaningful link and discovery on many levels. DNA is the only indestructible medium for planting the code of life, so long as life continues to exist. Placing a message there makes perfect sense. It directly links faith to mathematical patterns and code messages that link humans to the universe beyond Earth.

Seeking an explanation for the grand mysteries of why our earth and the universe looks and behaves the way it does and trying to predict what the future might have in store for us continues to be a never-ending quest. That quest includes establishing open communication with ETs. Our search for answers must be in the form of mathematics. That makes at least some sense to us. Mathematical equations are provable and therefore defensible. By mathematically modeling the natural world that we can see, smell, and touch into a code of numbers, we have identified many things that appear to govern and therefore make predictable many of the behaviors of our earthly world and the universe beyond it. *Angel Communication Code* is not about mathematics; however, there are some fundamental mathematical concepts that must be discussed in a general way, such that the road we are traveling in search of the code for ET communication and the involvement of angels makes sense and has a sound basis. This project to communicate with ETs is not a fool's errand. It is critical to the future of humanity. That is why billions of dollars have been, and continue to be, invested in the search year after year, with no end in sight.

Understanding at least some pieces to nature's code has allowed us to do more than just survive. It has allowed us to change our surroundings and to create new technologies that have resulted in the world that we know and live in today. We have used mathematics to help us create and understand things that make our lives better. Mathematics is the universal code that has allowed us to work with and often control our natural world with the intention of making our lives easier, safer, healthier, and generally more comfortable. It is important to recognize that we have made mistakes using this knowledge that have resulted in some very bad things as well.

The entire premise of this ET communication project is that the code that leads us to ET communication begins as a relatively simple mathematical code laid out for us either by ETs or through angels, or both. This is all about going back to basics and rethinking everything in our historic pursuit of ET communication. This is not about rehashing the question of *if* ETs exist. We must acknowledge that we are past that realization. We must now concede that what we have done in the past to establish ET communication has apparently not been successful. We must move on from using the same processes for establishing ET communication and expecting a different result. That is one part of the process that has become a fool's errand.

---

5 Joseph M. Jordan, "How Many Times Is Yahweh Mentioned in the Bible (What Does the Bible Say about Yahweh)?," Christian Faith Guide, 2023.

6 Wikipedia, s.v. "Yahweh."

The simplest question to consider is why ETs don't just make themselves readily available for communication with us. Why must there be a code for finding a message, like some sort of treasure hunt? The considerations are these:

1. Extraterrestrials were no doubt here on earth in ancient human history (or earlier).
2. ETs interacted with humans and provided knowledge and technology way beyond human capability to conceive of or manufacture in ancient times.
3. ETs have largely ceased making direct contact with us and likely have been observing our development for thousands of years.
4. ETs intend to reconnect when humans have advanced sufficiently to be successfully introduced to the community of the universe beyond Earth.
5. Before ETs disconnected, they laid out a map and left clues to a code or patterns for us to discover and decipher when we become sufficiently advanced, not only to find the clues, but also to understand them and use them to make contact with ETs.
6. When humans achieve a sufficient level of advancement, ETs will appropriately respond to our communication efforts.

The ET communication issue is much deeper than pure mathematics and science. It is also about the ethics of interference. A good example of this is the concept of the "prime directive" of the fictional television and motion picture series *Star Trek*. The prime directive prohibits interference (but not contact) with the other cultures and civilizations that Starfleet members encounter in their exploration of the universe. The mission of Starfleet is to "seek out new life and new civilizations." The prime directive's intent is to forbid interference with the natural development of other civilizations, regardless of their level of social or technological advancement. The executive summary of the order given in the 1968 episode "Bread and Circuses" is "no identification of self or mission, and no interference with the social development of said planet, and no references to space, other worlds, or advanced civilizations."

Starfleet officers take an oath to uphold the prime directive even if it means sacrificing their own lives or the lives of their crew, yet frequently they violate this directive when faced with circumstances where strict adherence to t the prime directive becomes inadvisable, the directive itself then becoming subject to interpretation. The directive and the mission seem to be at odds with each other anytime there is interaction between ET life and Starfleet. Unless they just observe from afar, the Starfleet crew are interacting, and because of this, they have influenced the other civilization just by making their presence known. In addition, Starfleet often leaves advanced knowledge and sometimes technology behind after they disengage from their "new life and new civilization" contact. The things they leave behind are intended to improve (if not save) the lives of those of this new civilization and help them advance.

This is consistent with the real-life situation between ancient human and extraterrestrial interactions in our own history. Sharing the universe becomes a genuine ethical question in this light. The prime directive demonstrates the conflict between "consequentialist" commitments to reducing harm and a "Kantian" commitment to respect the autonomy of others.

Consequentialism is the ethical position that considers an action to be either right or wrong judging by the consequence of that action. For example, most people would agree that lying is wrong; however, if telling a lie could save a person's life, consequentialism says lying is the right thing to do. There are many situations where this question comes into play.

Then there is the Kantian position. Immanuel Kant believed that the ultimate principle of morality was one of practical rationality that he defined as the categorical imperative (CI). CI is described as an objective that is rationally necessary and as an unconditional principle that people must follow despite any instinctive desires we may have to the contrary. Kant argued that all specific moral requirements are justified by the CI principle. This therefore means that all immoral actions are irrational because they violate the CI. In simpler terms, it means that humans are required never to treat others as merely a means to an end, regardless of what that end might be. Kant used the example of lying as an application of his ethics: "because there is a perfect duty to tell the truth, we must never lie, even if it seems that lying would bring about better consequences than telling the truth. The duty not to lie always holds true."[7]

Built into the fictional prime directive is an assumption that cultures will be better off if left to their own devices, whether those devices are social practices or technologies. Interference, even if well-intentioned, is assumed to disrupt outcomes in unforeseeable ways. It can be likened to the concept of going back in time and changing an event, which would alter in unpredictable ways a future that already happened once. If a culture must contend with the consequences of poor choices, then the people of that culture would be better off if those consequences were the result of their own free choices. It is a moral belief that it is right to respect a civilization's values, beliefs, and practices rather than imposing what we perceive as better ones upon it. A good example would be how European settlers of North America tried to "civilize" the Indigenous population.

A directive of noninterference is an ethical philosophy. The goal would be to understand how a culture and its citizens behave while the observers remain anonymous. To intervene is to influence natural behaviors, including the development of indigenous technologies that result. Even just observing a civilization can interfere with it, depending on your method of observation. People studying animal behavior in the field, if noticed by the animals being studied, can and probably do affect the natural behavior of those animals, depending on the animal (ant versus primate, e.g.). If the genuine goal is not to change the natural behavior or the development of ET civilizations at any cost, then it is probably best to not to explore new worlds at all, assuming observation (and not conquest) is one's only purpose. Based on the ethical and philosophical options, one must conclude that the ET position was one of interference in antiquity and that ETs continue this position of interference in modern times. It is their intent to influence the behavior and technological development of human beings. Our history would be much different if ETs had not done that in ancient times. In addition, our behaviors surrounding this matter today are influenced by the fact that we know that ETs exist and are out there somewhere. Some will argue that ET influence goes as far as human DNA manipulation. It is entirely possible some of that ancient ET influence on the earth will have more of an unexpected side effect and not one that was predicted or intended.

Sharing the universe is a different paradigm from treating any ET culture purely as a subject of study. Sharing a universe with other beings puts you in a different relationship with those beings. Sharing a universe with other beings requires reciprocity, even if respective technological achievements are not equal. Perhaps this is what allows the concept of a prime directive to be subject to interpretation. Sharing the universe can be considered an intervention with the true goal of finding a way to share it with each entity that is in existence on like and equal terms.[8] It all starts with the first step, which is to establish intelligent and open two-way

7 *Stanford Encyclopedia of Philosophy*, ed. Edward N. Zalta (Stanford, CA: Stanford University, Winter 2021), "Kant's Moral Philosophy," Robert Johnson and Adam Cureton.

8 Janet D. Stemwedel, "The Philosophy of Star Trek: Is the Prime Directive Ethical?," *Forbes*, August 20, 2015.

communication, for better or for worse. Either way it is going to happen, and according to all predictions, be they based in science or theology, it is going to happen soon.

It is reasonable to expect that any ET communication code laid out for us to discover would be simple enough for us to find it if only we would look in the right places. It must surely be linked to the universal code of mathematics and prime numbers. More specifically, the code must be linked to the first three prime numbers 2, 3, and 5. These prime numbers are the key. As is the case with any coded message, one needs a key to unlock it. The code must be discoverable using the mathematical keys, but not obvious in the sense that one is merely handed the key and simply inserts that key into the lock and twists it.

Long before the contemporary science of mathematics, people of ancient cultures sought clues and patterns left by the animals they either hunted for food or hid from as a threat to their lives. Footprints, sounds, odors, and color patterns are all clues as to where a prey or predator animal is at the moment, what its intentions might be (mating or territorial defense), where it has been, or where it might be going next. These are clues that can be identified and used by predators to help them get to their next meal. It is the very definition of hunting. Animals, including humans, will leave clues intentionally to alert all others that they are in the neighborhood as messages to potential mates or rivals. This is a relatively simple concept describing something that goes on to this day among all creatures of the earth. The first thing humans did when we first walked on the moon was to plant a flag in the ground to let future people or ETs know intelligent life had been there. This is an instinctual animal behavior that has become much more complex over time. It becomes more complicated when we consider humans trying to hide messages (cover their tracks) in patterns and codes intended for deciphering only by the ones with whom they wish to share the information and not by all. It also becomes very complex when we try to identify patterns that model the universe and the dynamics of inanimate objects in motion or predict the behavior of biological entities.

Going back to basics begins with the fundamental theorem of arithmetic or "unique factorization" that has its roots in ancient Greek arithmetic. The first credible statement of this theorem was offered by Carl Friedrich Gauss in 1801 in his book *Disquisitiones arithmeticae* (Arithmetical Investigations [Latin]).

The statement of the fundamental theorem of arithmetic is: "every composite number can be factorized as a product of primes, and this factorization is unique, apart from the order in which the prime factors occur." For example, the prime factorization of 240 would be:

$$240 = 2 \times 2 \times 2 \times 2 \times 3 \times 5$$

This theorem further tells us that this factorization must be unique. That is, there must be no other way to express 240 as a product of primes. We can change the order in which the prime factors occur. For example, the prime factorization can be written as $240 = 3^1 \times 2^4 \times 5^1$ or $3^1 \times 2^2 \times 5^1 \times 2^2$ etc., but the set of prime factors (and the number of times each factor occurs) is unique. That is, 240 can have only one possible prime factorization, with three factors, which are one factor of two, that is, 2^4; one factor of 3, that is, 3^1; and 1 factor of 5, that is, 5^1.[9] The point of this exercise is to demonstrate the importance of prime numbers 2, 3, and 5 in the universe as a logical tool for seeking an ET communication code.

9 "The Fundamental Theorem of Arithmetic," CueMath.

The prime factorization of our subject 2, 3, 5 primes expressed as the whole number 235 (5×47) is a prime number. That is interesting with respect to all the things we are discussing in *Angel Communication Code* because the Bible credits Jesus with performing forty-seven miracles.

Gauss proved the existence of a prime decomposition in 1982, which offered everything necessary to prove its uniqueness.[10] Every number is built by multiplying prime numbers together, and from numbers you get mathematics, and from mathematics you get the entire universe of science. What you do not get is an imperial belief and faith in a Creator, God, or angels. People of faith (almost any faith) believe that God or some other creator made it all possible in the first place. Mathematics was not invented but, rather, was and continues to be a process of discovery.

Prime numbers are the DNA of the arithmetic. Their existence is fundamental, yet prime numbers remain one of the greatest enigmas in mathematics. Nobody has ever been able to identify a reliable code that will predict where the next prime number will be found. We know prime numbers go on forever, but finding the pattern has been one of the biggest mysteries in mathematics. A million-dollar prize has been offered by the Clay Mathematics Institute of Cambridge, Massachusetts, to anyone who can reveal the pattern of prime numbers.[11]

The numbers 2, 3, and 5 are the first numbers greater than the number 1, and the first three prime numbers in the Fibonacci number sequence. This is much less simple, but clearly our root code numbers point us in this direction. The Fibonacci sequence is not discussed in the *Extraterrestrial Communication Code* book, so we will briefly touch upon it here.

The Fibonacci sequence of numbers is named after Leonardo Pisano, whose nickname was Fibonacci. He was an Italian mathematician who lived from AD 1170 to 1250. The Fibonacci numbers are a sequence where each number is the sum of the two preceding ones, as follows:

0, 1, 1, 2, 3, 5, 8, 13, 21, 34, 55, 89, 144, and so on

The Fibonacci numbers were first described in Indian mathematics as early as 200 BC by a man called Pingala. He was trying to determine mathematical patterns within Sanskrit poetry formed from syllables of two lengths. It was Fibonacci who introduced the sequence to Western European mathematics in his 1202 book *Liber abaci* (Book of calculation).

Fibonacci numbers appear so often in mathematics and in nature that there is an entire journal dedicated to their study: the *Fibonacci Quarterly*, which is the official publication of the Fibonacci Association. The following is taken from their website:

> As the primary publication of the Fibonacci Association, the *Fibonacci Quarterly* provides the focus for worldwide interest in the Fibonacci number sequence and related mathematics. New results, research proposals, challenging problems, and new proofs of known relationships are encouraged. The *Quarterly* seeks intelligible, well-motivated, university-level articles.

---

10 Aryan Thakur, "Fundamental Theorem of Arithmetic," Protons Talk, protonstalk.com/basic-math/fundamental-theorem-of-arithmetic, accessed November 24, 2023.

11 "Nature's Hidden Prime Number Code," *BBC News Magazine*, July 27, 2011.

Illustrations and tables should be included to the extent that they clarify main ideas of the text. A well-developed list of references is required.

Common applications of these numbers include:

- computer algorithms such as the Fibonacci search technique
- the Fibonacci heap data structure
- graphs called Fibonacci cubes that are used for parallel and distributed systems. In nature they are known to describe (among other things):
    - The arrangement of seeds on a sunflower
    - the configuration of leaves on a stem
    - the sprouts of a pineapple
    - the uncurling of a fern
    - the bracts on a seed-bearing pinecone
    - the flowering of an artichoke.

Fibonacci numbers define a perfect spiral and are of interest to biologists and physicists because they are frequently observed in various natural objects and phenomena. The branching patterns in trees and leaves, for example, and the distribution of seeds in a raspberry reflect the Fibonacci sequence.[12] The *Mona Lisa* by Da Vinci fits the Fibonacci model perfectly, as do other classical works of art. Da Vinci is well-known for building secret symbols and codes into many of his works.

Fibunacci Number	Devine Ratio
1	N/A
1	1.000
2	2.000
3	1.500
5	1.667
8	1.600
13	1.625
21	1.615
34	1.619
55	1.618
89	1.618
144	1.618
233	1.618
377	1.618
610	1.618
987	1.618

The Fibonacci sequence is directly linked to the golden or "divine" ratio, 1:1.618, which is found all over the natural world. The golden ratio is derived by dividing each number of the Fibonacci series by its immediate predecessor. In mathematical terms, if F (n) describes the nth Fibonacci number, then the quotient F (n) / F (n–1) will approach the limit of 1.618 for increasingly high values of n. This limit is the divine ratio.[13] A table of these values is shown to the right. Notice the divine ratio begins at Fibonacci number 55; the number 5 is important to the ET code derivation we are developing in *Angel Communication Code*. Also, note that the divine ratio repeats to three decimal places.

It is one of the most famous ratios in mathematics, art, and design going all the way back to the ancient Greeks. It is typically written as the Greek letter phi. DNA is shaped around this ratio. The divine proportion is said to be the only mathematical ratio that the universe needs to design all life.

The first known mention of the divine or golden ratio is from around 300 BCE in Euclid's *Elements*, the classical Greek work on mathematics and geometry. Euclid and other early mathematicians such as Pythagoras recognized the proportion but didn't call it the golden ratio. It wasn't until much later that the proportion would take on its persona. In 1509, Italian mathematician Luca Pacioli published the book

---

12 Wikipedia, s.v. "Fibonacci Number (and References)," May 30, 2022.

13 "Fibonacci and Golden Ratio," Let's Talk Science, May 24, 2022.

entitled *De divina proportione* (The Divine Proportion). Along with works by Leonardo Da Vinci, it promoted the ratio as divinely inspired simplicity and orderliness. The ratio is said to be a mathematical construct of beauty that is naturally pleasing to the eye.[14]

In the natural world, the divine ratio controls objects as large as galaxies and as small as DNA. The divine ratio represents perfect harmony, or the most attractive proportion in almost all things. Adrian Bejan, a Duke University engineer, has found it to be a compelling ratio for a single law of nature's design. In numerous papers and books, Bejan has demonstrated that the constructal law shows how all shape and structure in nature arises to facilitate flow. Bejan discusses how his constructal law provides a greater scientific context for nature's efficiencies such as the divine ratio. Designs in nature are completed utilizing the minimum of resources and energy required. The divine ratio models the natural movement of all known energy in the universe.[15]

The divine ratio creates a form that can increase in size indefinitely without altering its shape. In this way, across size scales from very small to very large, the same form arises repeatedly. The divine ratio enables flow optimization because it follows a path of least resistance, so that a maximum result can be achieved with the least amount of effort. The divine ratio explains why falcons fly in a spiral path when approaching their prey. For falcons, this spiral is the energy-efficient flight path of least resistance. Fibonacci spirals are the configuration of least energy, and experimental results provide a vivid demonstration of this energy principle in phyllotaxis, the arrangement of leaves on a plant stem (Li, Ji, and Cao 2007).

The constructal law shows there is an actual purpose to all life as well as a meaning to individual life. It reconciles modern science with ancient scriptures and spiritual writings. It suggests we consider that everything in the universe is ultimately made up of energy. One of the most important principles of energy is that it doesn't like differences and it works out ways to reduce and balance them in an effort to achieve equilibrium. This is why energy flows from a place where it is more concentrated to a place where it is less concentrated.

While the twenty-five chemicals that make up our human bodies are the same for everyone, the way energy is mixed with them is different for each individual. We all have bodies with similar brains with a similar number of nerves in each, but the way those nerves are connected is different in each of us. When your energies are flowing together, focused in one direction, you may experience what psychological literature defines as peak flow or peak experiences. This sensation is like being carried along by the flow of an effortless current of some type.[16]

Science is forever trying to model the universe using mathematical equations. Using numbers to model the physical world is essentially what science is all about. Our comprehension of things in the universe as those things relate to numbers is limited only by the ability of our human minds to comprehend complex concepts. For example, at any given instant in time, there is a finite number of grains of sand on the world's beaches, there is a finite number of drops of water in the world's oceans, and there is a finite number of oxygen molecules in Earth's atmosphere. Those numbers change, presumably in a pattern that theoretically

---

[14] Jacob Obermiller and Sara Berndt, "An Introduction to the Golden Ratio," Adobe, https://www.adobe.com/creativecloud/design/discover/golden-ratio.html, accessed November 24, 2023.

[15] A. Bejan, "The Golden Ratio Predicted: Vision, Cognition, and Locomotion as a Single Design in Nature," *International Journal of Design & Nature and Ecodynamics* (2009).

[16] "The Purpose of Life and Golden Ratio Explained," Fibonacci Life-Chart, April 21, 2017.

could be mathematically modeled for every second in time as the earth turns and the universe does whatever it is doing. Can we truly comprehend concepts and numbers that large, and more importantly, do those concepts and numbers mean anything useful to us? They must mean something, but they probably have no practical application for humans. Those sorts of numbers are just too big to grasp or accurately quantify. The same is true about the concepts and numbers we toss around when discussing the universe and the heavens. What does it really mean to say something is a billion light-years away from Earth? These sorts of numbers are enormous concepts coming from Earth, which is less than a fraction of a pinprick in the universe. The universe is too enormous for most, if not all, humans to truly grasp. So why does ET communication with beings that could be light-years away from Earth even matter?

It matters because we know with a high degree of certainty at this point in time that ETs have been coming to Earth for thousands of years and continue to come here to this day. They are somehow able to close the time and distance gap in a way humans have not figured out yet. They are or can be an arm's length away, not billions of light-years away. It matters because if they can get to us here on Earth, then we will inevitably be able to get to them someday on their planet, either under our own power or by hitching a ride with them, or by way of some sort of dimension of consciousness. And if this is the case, then we need to be able to communicate with them. We need to know what their intentions might be because it's a matter of our survival as a human race.

Since the days of Noah after the Flood, it seems that open and direct communication with the divine (other than prayer) and ETs has all but vanished. Note that the abduction of humans by ETs is not two-way intelligent communication. Abductions are a very different matter entirely—and scary. We have tried many times to instigate intelligent two-way communication with ETs, but those attempts have thus far failed to the best of our knowledge. We explore a new means and method of establishing ET communication in *Angel Communication Code*, versus being close-minded and continuing to do what we know has not worked. Breaking free from unsuccessful ideas and failed methods starts by thinking differently and going back to basics—back to the drawing board, where we may find that in our intellectually myopic approach to the problem, we have overlooked the simple and obvious clues and answers. There is no simple answer to the question of two-way ET communication, but we must understand that if we think like a hammer, then everything in the universe will look to us like a nail. This thought process opens the door to a discussion about chaos theory.

Edward Lorenz, a MIT meteorologist who tried to explain why it is so hard to make good weather forecasts, wound up unleashing the scientific revolution called chaos theory. Also a professor at MIT, Lorenz was the first to recognize what is now called chaotic behavior in the mathematical modeling of weather systems. In the early 1960s, Lorenz realized that small differences in a dynamic system such as the atmosphere, or a model of the atmosphere, could trigger vast and often unsuspected results. These observations ultimately led him to formulate what became known as "the butterfly effect," a term that grew out of an academic paper he presented in 1972 entitled "Predictability: Does the Flap of a Butterfly's Wings in Brazil Set Off a Tornado in Texas?"

Lorenz's early insights marked the beginning of a new field of study that impacted not just the field of mathematics but also virtually every branch of science: biological, physical, and social. In meteorology, it led to the conclusion that it may be fundamentally impossible to predict weather beyond two or three weeks with a reasonable degree of accuracy. Some scientists have since asserted that the twentieth century will be remembered for three scientific revolutions: the theory of relativity, quantum mechanics, and chaos theory.[17]

Chaos theory attempts to model a set of data focused on the effect of forces on the motion of objects or "dynamics." The root of most, if not all, theories of mechanics is based on Isaac Newton's laws or Newtonian physics.

Isaac Newton is recognized in the scientific community as the founder of classical mechanics. This is based on his belief that space is separate from the body and that time passes uniformly whether anything happens in the world or not. It is a complicated concept coming from the person who is fabled to be the one who had an apple drop on his head, resulting in the identification of gravity. Newton spoke in terms of absolute space and absolute time. From ancient days and all the way into the eighteenth century, competing opinions disputed that space and time are real entities. The opponents to Newton believed that the world is a purely material entity. They maintained that the existence of empty space is a physical impossibility. In addition, they maintained that there can be no passage of time without change occurring somewhere. Time is a measure of cycles of change within the world. Newton defined the true motion of a body to be its motion through absolute space. The nonbelievers thought that the concept of true motion could be measured in terms of relative motions and their causes. The difficulty of doing this made a strong argument for Newton's idea of existence of absolute space.[18]

The notion that everything in the universe is linked to a specific cause or a variety of causes has been debated for centuries, particularly during the seventeenth century. This was when contemporary astronomers of the time became capable of predicting the trajectories of planets.[19] That being said, the ability for humans to track the movements of celestial bodies and build structures to align with these movements has been going on since the beginning of civilization. The point is that humans have always been in search of patterns and codes to describe the natural world that can be seen through a microscope, with their own unaided eyes, or through a telescope. When something that is unknown and perhaps a little frightening becomes less unknown and reasonably predictable, we feel much safer.

The method by which humans strive to understand things in the universe is to create a mathematical model of the concept or an observable phenomenon in a way that can be managed within the limits of our ability to think, understand, quantify, and use numbers to model the unknown. This is true for earthly things such as the quantification of grains of sand, drops of water, and molecules of air. It is also true when we consider the concept of time, the vastness of the universe, and the notion of not being alone in the universe and establishing two-way communication with ETs.

Now consider the more mystical indicators as to what is out there in the universe that was conceived, created, and planted by humans (or others) even before there was a human understanding of higher-level mathematics. In all forms of religion and mythology there are relevant number patterns that are not so obvious until one does a bit of digging. Religious texts and folklore are not authored or told to someone by the subject "god," but rather they are authored and told by humans for the purpose of communicating the

---

17 "Edward Lorenz, Father of Chaos Theory and Butterfly Effect, Dies at 90," MIT News, April 16, 2008.

18 *Stanford Encyclopedia of Philosophy*, "Newton's Views on Space, Time, and Motion."

19 Christian Oestreicher, "A History of Chaos Theory," *Dialogues in Clinical Neuroscience* 9, no. 3 (2007): 279–89, https://doi.org/10.31887/DCNS.2007.9.3/coestreicher.

story of things not of this earth. Buried within these human-to-human-communicated stories are simple numerical clues that mean something more. Across the board, the ancient numerical clue seeds planted by ETs, angels, or ancient mystics for future human deciphering often seem to involve prime numbers 2, 3, 5, and 7. The communication code keys can be found in the manipulation of these prime numbers.

Long before there was the written Bible, there was Norse mythology, which today is often dismissed as pure fiction. Religious practices represented an essential element of early Germanic culture. The roots of Norse mythology grew out of Germanic paganism, which included a variety of religious practices of the Germanic peoples from the Iron Age (between 1200 BC and 600 BC) to the Christianity of the Middle Ages (500 to 1400–1500 CE).

Norse mythology is made up of two main groups of beings that were considered to be gods to be revered in the pantheon (temple of all gods). These two groups were identified as the Aesir and the Vanir, which represent different forces in the cosmos. The Aesir are the gods of morality, idealism, manhood, motherhood, and valor. The Vanir represent the forces of nature responsible for fertility, growth, sensuality, and pleasure. Another, lesser group was known as the Jötnar, who are not gods but are destructive forces that represent the dangerous and chaotic aspects of the cosmos.[20]

There are numerical clues to a larger story buried within the folklore. Norse mythology paints the picture of indigenous pre-Christian religion, beliefs, and legends believed by Scandinavian peoples. Iceland is where we believe most of the written sources for Norse mythology originated. These stories were passed down in the form of poetry until the period between the eleventh and eighteenth centuries, when the Eddas and other medieval texts were written. Norse mythology is considered the best-preserved version of the even more ancient Germanic paganism, which is closely linked to Anglo-Saxon mythology and folklore.[21] Woven through this mythology are links to our subject prime numbers, presented for a reason that we need to pay attention to as germane to the subject of *Angel Communication Code*.

The number 3 is the most important, and it turns up all in many places, including the Bible and Norse mythology. Some (but not all) important Norse references to the number 3 are as follows:[22]

- There were three original beings: the primordial cow Audhumla; Ymir, the first giant; and Búri, the first god and the grandfather of Odin.
- For three days Audhumla licked the ice of Ginnungagap until Búri was freed.
- Ymir had three direct offspring: a boy and girl who grew from beneath his arms, and a six-headed son who sprang from the coupling of his feet.
- There were three generations of giants before the race was destroyed by the deluge of Ymir's blood, after which time his grandson Bergelmir became the progenitor of a new line.
- The heart of the giant Hrungnir was triangular and made of stone.
- There are three named Norns—three wisewomen thread spinners—who determined every allotted life span. One spun out the thread of each life; another measured its length; and the third decided when the thread should be snapped.

20 Wulf Willelmson, "Angels and Demons in Teutonic Mythology," Creed of Caledon, March 16, 2017.

21 Norman, "The Origins of the Norse Mythology," TheNorseGods.com, November 20, 2013.

22 Wikipedia, s.v. "Numbers in Norse Mythology."

- Odin had two brothers, numbering three sons of Borr who created the world and gave life to the first human beings.
- Odin is the ruler of the third generation of gods as the son of Borr and grandson of Búri.
- Yggdrasil, the "World Tree," has three roots. Under the three roots are three sacred wells, one for each root.
- Odin endured three hardships upon the World Tree in his quest for the runes: he hanged himself, wounded himself with a spear, and suffered from hunger and thirst.
- Loki sired three monstrous children by the giantess Angrboda: the wolf Fenrir, Jörmungandr the World Serpent, and Hel, a female being who is said to preside over an underworld realm, also called Hel.
- There are three main events leading up to Ragnarök itself: the birth of Loki's three monstrous children, the death of Baldr and subsequent punishment of Loki, and the onset of Fimbulwinter, that is, three hard winters without separation by summers.
- The wolf Fenrir was bound by three fetters, Loeding, Drómi, and Gleipnir, of which only the last held him.
- Loki is bound with three bonds made from the entrails of his son through holes in three upright slabs of rock: the first under his shoulders, the second under his loins, and the third under the backs of his knees.
- In the poem "Völuspá" from the *Poetic Edda*, the description of the monstrous hound Garmr's howl is repeated three times.
- In Völuspá, the gods burn Gullveig three times, and three times she is reborn.
- During the onset of Ragnarök, three cocks will begin to crow, heralding the final conflict: Gullinkambi for the gods, Fjalar for the giants, and an unnamed third for the dead.
- Bifröst the rainbow bridge has three colors. It also has two other names, Ásbrú and Bilröst, and thus it has three names.
- Heimdall has three special powers in his role as guardian of the rainbow bridge. He needs less sleep than a bird, can see at night for a hundred leagues, and is able to hear grass growing on the earth.
- Odin has three special possessions: his spear Gungnir, his golden ring Draupnir, and his eight-legged horse Sleipnir.
- Thor has three main weapons for use against the giants: his hammer Mjolnir, a magical belt that doubles his strength, and a pair of iron gauntlets that allow him to wield the hammer.
- Freyr has three magical items, including the ship *Skidbladnir*, his boar Gullinbursti, and a sword with the ability to fight on its own.
- Freyja has three special artifacts, including the priceless necklace Brisingamen, a cloak that allows her to assume the form of a falcon, and a chariot drawn by a pair of great cats.
- In the stronghold of the giant Útgarda-Loki, Thor drank three drafts from a horn during a drinking contest but gave up when he was unable to empty the horn of its contents. This was one of three tasks he attempted during his stay. The other two were to lift a cat (he made it lift a paw, leaving three on ground) and to defeat an old woman. It is later revealed that the horn was connected to the sea (which he leveled down by three fingers), the cat was the World Serpent, and the old woman was old age itself. Before this, Thor and his companions had met the giant, who was under the assumed name Skrýmir, in the forest outside the castle. When Skrýmir had gone to sleep during their journey together, Thor became annoyed by his loud snoring and struck at him three times with his hammer, but in each case the blow was misdirected through magic and illusion.
- The builder of the walls of Asgard offered to build them in three seasons in return for three prizes: the sun, the moon, and the hand of Freyja in marriage.

- Odin spent three nights with the giantess Gunnlod to obtain the mead of poetry. She then allowed him to take three drinks of the mead, one from each of three vessels.
- There were three statues of Odin, Thor, and Freyr in the temple at Uppsala.
- Three of Odin's sons remain after Ragarök: Vidar, Baldr, and Hǫðr.

There is surely a reason why Norse mythologies point us so deliberately, frequently, and obviously to the number 3, as do several other important historical references and ancient texts.

Time moves forward into Christianity, and we consider the contribution of angels in the search for clues to an ET communication code. As with Norse mythology, there are no mathematical models quantifying the vastness of God and His angels or that predict when a miracle might occur. The mysteries of faith do not adhere to any mathematical laws. They have no mathematical boundaries that we have identified.

Searching for prediction models is essentially searching for a code. Humans do not invent these models, but rather the models were always there and humans just figured out how to express a given set of observations using mathematical models. Isaac Newton, for example, did not invent or discover gravity. He just observed it, then identified the mathematical model that could predict the dynamic impact of gravity on animate objects. Gravity was always there. Such is the case with all the laws of physics.

Consider the numerous megalithic structures we presume were constructed by ancient civilizations all over the world centuries before Newton's time. Either those civilizations had a very good grasp of how to work with or around gravity or they had help from ET beings that did. The building blocks were cut with incredible precision. The ancients then would have had to overcome gravity to lift multiton stones, move them, and transport them for miles across difficult terrain. Based on what we know about ancient civilizations and their level of technological advancement in general, it seems obvious that they had help from a more advanced entity not of this earth. It becomes even more obvious when we consider the similarity of size, shape, and celestial orientation of megalithic structures located in all corners of the globe. They were constructed at a time when these civilizations did not even know of each other's existence to the best of our current knowledge. To put this in context on a geological timescale, the historical record tells us that it was not that long ago when Europeans figured out that North America and South America even existed. Ancient civilizations of the world existed with seemingly no knowledge of the existence of other civilizations on the other side of the world, yet the scale, geometry, architecture, and alignment with celestial bodies of their megalithic structures is very similar, if not identical in many cases. Not one of these civilizations had the tools or technologies to construct these enormous structures and construct them with such precise geometry and alignments with celestial bodies, and also with correct longitudinal and latitudinal alignment with each other on opposite sides of the world. Not all these relationships can be a giant coincidence. The more logical and likely explanation is that there is an extraterrestrial connection. This phenomenon is a perfect example of a circumstance where the principle of Occam's razor applies.

Occam's (or Ockham's) razor is the term used to reference the law of economy or the law of parsimony, stated by the Scholastic philosopher William of Ockham (1285–1347/49) as "pluralitas non est ponenda sine necessitate," or "plurality should not be posited without necessity." The basic philosophy is that given two competing theories, the simpler explanation is to be preferred. The principle is also expressed as "Entities are not to be multiplied beyond necessity."

The principle was introduced before Ockham by Durandus of Saint-Pourçain, who was a French Dominican theologian and philosopher. He applied the principle to explain that "abstraction is the apprehension of some real entity, such as an Aristotelian cognitive species, an active intellect, or a disposition, all of which he spurned as unnecessary." In the field of science, a fourteenth-century French physicist named Nicole d'Oresme applied the principle (as did Galileo) in defending the simplest hypothesis of the heavens. Application of the principle became so frequent that it became known as "Occam's razor."[23]

So, which makes more sense? Did ancient civilizations construct their megalithic structures with hand tools, slave labor, and beasts of burden? Did they somehow figure out how to cut stone with perfect precision and move and lift hundred-ton (or heavier) boulders cross-country to the specific locations of construction that line up with stars, earth lines, and other structures on the other side of the world, structures that have been standing for thousands of years? On the other hand, did they have a wee bit of help from entities not of this earth that possessed the technology to travel to Earth from another part of the universe? There is not much debate at this point that ETs are out there and have been to Earth. This idea is much more defensible today than it would have been as little as a hundred years ago.

There is a tremendous difference between:

1. identifying and modeling patterns to describe the naturally occurring dynamics of animate objects in the universe, and
2. developing and interpreting codes intended to be understood by only those who know how to decode the information as presented in current time, and
3. discovering codes and patterns about ET communication that were intended to be discovered and interpreted when the time was right with respect to future intellectual and technological development of the message receiver—humans.

Creating and deciphering codes is the science of cryptology, the study of securing communications from unauthorized recipients. One who either creates or conversely "cracks" or deciphers a coded message is a cryptographer. Encryptions (coded messages) take an original message and convert it into ciphertext, which is not understandable at face value. A predetermined key to the encryption allows the receiver to decrypt (decode) the message, thus ensuring that only those who know the key can read and understand the message, unless the code is somehow cracked by uninvited recipients.

Cryptography has four primary objectives:

1. Confidentiality
   Confidentiality ensures that only the intended recipient can decrypt the message and understand its contents.

2. Nonrepudiation
   Nonrepudiation means the sender of the message cannot backtrack in the future and deny his or her reasons for sending or creating the message.

3. Integrity
   Integrity focuses on the ability to be certain that the information contained within the message cannot be modified while in storage or during transmission.

---

[23] *Encyclopedia Britannica*, s.v. "Occam's Razor (Philosophy)," Brian Duignan.

4. Authenticity
   Authenticity ensures the sender and recipient can verify each other's identities.

The word *cipher* is synonymous with the word *code*. Both words mean a set of steps that encrypt a message. The earliest-known cipher was called the "scytale." The scytale (or skytale) was a cylinder with a strip of parchment wrapped around it on which was written a message. It was used by the ancient Greeks and Spartans to communicate secretly during military campaigns.[24]

A cipher can be as simple as the decoder pin Ralphie used in the motion picture *A Christmas Story* to decode Little Orphan Annie's message after he received the decoder pin in the mail and the key was given over the radio by host Pierre Andre. Andre announced to the audience, "Remember, only members of Annie's secret society can decode the message."

In today's world of cybersecurity, a cipher needs to be extremely complex, something way beyond Ralphie's decoder pin, something that only sophisticated computer software programs can unravel. Cybersecurity has evolved into a very complex game of digital cat and mouse.

The encrypting procedure is dependent on the correct usage of the key. Without the correct usage of the key, it would be virtually impossible to decode an encrypted message. That is, after all, the fundamental purpose of an encryption: to conceal a message from unwanted eyes.

Ciphers were much easier to unravel at the beginning of their development as compared to modern cryptographic algorithms; however, they still used and continue to use keys and plaintext. In the modern world, encryption algorithms are far more sophisticated. Modern codes use numerous rounds of encryption ciphers to create the most secure transmission and storage of data. Today, there are techniques used that are said to be irreversible, which in theory protects the security of a message forever. Most of the ciphers and algorithms used in the early days of cryptography have been deciphered and the secret is no longer a secret.[25]

There are many examples of codes in recent history that were created by humans intended to conceal and deliver messages to only the people with whom the messenger intended to communicate. There are also many historic examples of the ancient use of encryptions.

One of the most historically significant encryption devices was known as the Enigma machine, a cipher machine developed and used in the early to middle twentieth century to protect commercial, diplomatic, and military communications. The Enigma was invented by a German engineer named Arthur Scherbius at the end of World War I. That fact was unknown to the general public until 2003. A paper by Karl de Leeuw was found that described Scherbius's work in detail. The German firm Scherbius & Ritter, cofounded by Scherbius, patented ideas for a cipher machine in 1918 and began marketing the finished product under the brand name "Enigma" in 1923. It was marketed to commercial markets but quickly caught on to government use as one might expect.

The word *enigma* is derived from Latin and Greek words that mean "to speak in riddles" (from Latin *aenigma*; from Greek *ainigma*). It applies to things, as well as to people who puzzle one's mind. Ancient megalithic

---

[24] "The Skytale: An Early Greek Cryptographic Device Used in Warfare," HistoryofInformation.com.

[25] "What Is Cryptography in Security? What Are the Different Types of Cryptography?," Encryption Consulting.

structures, quantum mechanics, the universe, ETs, and the late great physicist Stephen Hawking all qualify as being described as enigmas.[26]

During World War I, the United States was regularly deciphering coded messages sent by foreign diplomats. By deciphering diplomatic messages, the United States and Great Britain both knew what restrictions of weaponry would be acceptable to the Japanese in the peace talks following the war, and negotiators were able to do their work with this information in their back pockets. The effort to break Japanese diplomatic codes continued into the 1920s and 1930s under the direction of William Friedman, a Russian immigrant who was appointed chief cryptanalyst of the US Army Signal Intelligence Service (SIS) in 1922. In the late 1930s, SIS cryptanalysts succeeded in breaking the Purple Code, also designated AN-1. This was the primary cipher that Japan used to send diplomatic messages.

The Enigma machine was used extensively by Nazi Germany during World War II in all branches of the German military. It was considered so secure that it was used to encipher all top secret messages. The security of the Enigma machine codes was dependent upon machine settings that were generally changed daily, based on secret key lists distributed in advance. The receiving station would have to know and use the exact settings or "keys" set by the transmitting station to successfully decrypt a message.

While Nazi Germany introduced a series of improvements to the Enigma over the years, and while these improvements hampered decryption efforts, they did not prevent Poland from cracking the machine codes as early as December 1932 with the growing fear of a German invasion prior to the war. The Allies were able to decipher Enigma-encrypted messages, which were a major source of military intelligence. The Poles turned their information over to the British, who set up a secret code-breaking group known as Ultra. Because the Germans shared their encryption device with the Japanese, Ultra also contributed to Allied victories in the Pacific.[27] Cracking those codes most certainly shortened the duration of the war substantially and influenced its outcome.

On December 7, 1941, Japanese military forces attacked the United States naval fleet anchored at Pearl Harbor off the Hawaiian island of Oahu. The surprise attack was devastating to the US Navy. In the months that followed this attack, there were great concerns that Japan actually had the ability to launch such an invasion without our knowing it was coming in advance. Because the US Pacific Fleet was so badly crippled, the concern became that the Japanese navy could now fully control the Pacific Ocean, thereby cutting the United States off from critical supply lines and shipping lanes.

Over the next three and a half years, through several intense sea and island battles, US forces managed to stave off the Japanese Empire sufficiently enough to force them back to their own waters. They were able to do so because of the efforts of a great many men and women in battle and also thanks to the efforts of those who worked long and hard and in secrecy to successfully crack the codes that the Japanese were using to transmit messages revealing their plans. If the United States had not cracked the Japanese codes, they would not have been able properly prepared for engagement in Japanese offensives throughout the Pacific. They would not have known where the Japanese intended to strike next. The US did not have enough navy muscle left after Pearl Harbor to cover much more than one base at a time.

---

26 Merriam-Webster, s.v. "Enigma," https://www.merriam-webster.com/dictionary/enigma, accessed November 24, 2023.

27 *Encyclopedia Britannica*, s.v. "Enigma—German Code Device."

Many people consider that the most relevant and world-changing success that resulted from breaking the Japanese naval code was the Battle of Midway in June 1942. Through code decryption efforts by the United States, it was learned that the plan of Japanese commander Admiral Isoroku Yamamoto was to assemble an aircraft carrier task force and launch an air raid off the Aleutian Islands. It was their intent to lure the US Navy to Midway Island and engage in what they believed would be the decisive battle that would destroy what was left of the US fleet after Pearl Harbor. From decrypted messages, US naval commanders knew the general outlines of the plan, including the time it was planned. The decoded messages, however, did not specifically identify where the Japanese intended to wage this decisive battle. They identified the target only as "AF."

It was naval officer Joseph Rochefort who proposed a trap to determine what AF stood for. Rochefort was a cryptanalyst in the United States Navy's cryptographic and intelligence operations from 1925 to 1946, particularly recognized for his work in the Battle of Midway. The work of Rochefort and his team was critical to the US victory in the Pacific theater. He had a suspicion that the Japanese target could be Midway Island, so he arranged for US forces on Midway to send out a radio message saying that they were running short of fresh water. Rochefort and his team stood by to see if Japan would rise to the bait. Finally, codebreakers intercepted a Japanese message announcing that AF was running short of fresh water. Now knowing that the target was indeed Midway, the US Navy was ready and waiting on June 4, 1942, the date they already knew the attack was planned. After a hard-fought three-day battle, US fighter pilots sank all four Japanese aircraft carriers in Yamamoto's task force, thereby turning the tide of World War II in the Pacific theater.

Admiral Yamamoto was ultimately killed as a result of a decrypted message. Codebreakers discovered that he was scheduled to inspect a naval base on Bougainville in the Solomon Islands on April 18, 1943. US policymakers, however, were at first reluctant to use this information. They were concerned that using the information would alert the Japanese that their codes had been broken. In the end, however, the decision was made to assassinate Yamamoto. That morning, eighteen fighter pilots deployed from the US base at Guadalcanal, which was at the opposite end of the Solomon Island chain. They arrived on the target just as Yamamoto's plane was on the approach. Yamamoto was killed in the attack. The Japanese lost their most experienced and accomplished admiral, which was enough to break the back of Japanese morale.

To maintain the illusion that the United States had no advance knowledge of Yamamoto's plans and that the US fighters had arrived on the scene by an unfortunate twist of fate for the Japanese, the air force flew other patrols in the area, both before and after the attack. The Japanese did not catch on and, thus, did not change their encryption code. US intelligence went on to intercept and decipher thousands of Japanese messages.

In April of that year, decrypted messages revealed that Japanese forces were preparing for an assault on Port Moresby and an Australian base in New Guinea. US Admiral Chester Nimitz sailed his fleet into the Coral Sea between New Guinea and Australia as a result. While the ensuing two-day Battle of Coral Sea was considered a draw, US forces inflicted enough damage on the Japanese navy to force it to withdraw, giving the United States and Australia time to reinforce Allied defenses in New Guinea.[28] It can be argued that the ability of the United States to intercept, decipher, and respond to encrypted messages transmitted by their adversaries changed the course of modern history as we know it today.

---

[28] Michael J. O'Neil, "World War II, United States Breaking of Japanese Naval Codes," Encyclopedia.com.

There have been three major periods of cryptology that have been well-defined over the course of the history of cryptology. The first was manual cryptology. This began in ancient times and lasted all the way up to and through World War I. It was constrained by the limitations of what a code clerk or cipher had the ability to do using simplistic devices. Ciphers were limited to, at most, a few pages or a few thousand code characters. The principles for both cryptography and cryptanalysis were known; however, the security that could be achieved was always limited by what could be done manually.

The second phase used mechanical cryptography and began shortly after World War I with the Enigma machine. The third phase is digital and has only truly and effectively been around since the 1980s. The point is that the science of cryptology has been around for very long time to deliver messages over long distances that people wanted kept private should the message fall into unwanted hands.[29] What we are doing in *Angel Communication Code* with respect to identifying a coded message is not a new or unusual thing, but it is a different twist on an established thing.

Now consider the sciences of string theory and supersymmetry, which seek to decode the most fundamental workings of the universe. String theory, in particle physics, is a theory that attempts to merge quantum mechanics with Albert Einstein's general theory of relativity. The name *string theory* comes from the modeling of subatomic particles as tiny one-dimensional "stringlike" entities, rather than the more conventional approach in which they are modeled as zero-dimensional point particles. In the 1980s, physicists realized that string theory had the potential to incorporate all nature's forces. While string theory is still an exciting area of research that is undergoing rapid development, it remains primarily a mathematical construct because it has yet to make contact with experimental observations.[30]

A scientist named S. James Gates, a theoretical physicist at the University of Maryland at College Park, developed a theorem based on string theory that there is, in fact, computer code embedded in the fabric of reality, which would be very strong evidence pointing to a Creator. This is not the DNA coding discussed previously. This is the computer code that controls the universe, and it did not arise by chance. The idea is that there is no possible way computer code could have evolved into original existence on its own. Someone must have designed it and put it in place. Proof of the theorem would be hard evidence, more than just faith, that an intelligent being had had a hand in designing our natural world. Dr. Gates's findings are solid according to University of Maryland physicist Tristan Hubsch, who assures that the discovery of coding in supersymmetry is a rigorously proven theorem.

It all came to light when Gates was developing diagrams to represent equations in string theory. He noticed that these equations are indistinguishable from error-correcting coding found in browsers. Browser coding is a special kind of coding developed by a mathematician named Claude Shannon, a pioneer in information theory.

Just as browser coding obviously had to have been written by an intelligent mind, the same is true for coding found in nature. The theorem is not about literal computer coding such as found in C++ or Visual Basic. It is a type of math that's indistinguishable from Shannon coding, which is a mathematical way to represent information.

---

[29] *Encyclopedia Britannica*, s.v. "The History of Cryptology."

[30] *Encyclopedia Britannica*, s.v. "String Theory Physics," Brian Greene.

It is an interesting intellectual exercise to think about the way the universe might function being like a computer or quantum computer, but to suggest that we are living in a matrix-like simulation seems a bit of a stretch. In an interview with *Smithsonian Magazine*, Gates said, "I have never found a schism in my life between doing science and having religious beliefs …. I believe that both faith and science are essential for the survival of our species."[31] Creation and big bang theorists are beginning to concede to portions of each other's positions on this matter. The point is that there are areas of research going on in our world that are quite imaginative that make the subject of this ET and angel communication book not so extraordinary a concept.

In recent years, the need to protect digital messages and personal information with codes, usernames, and passwords has become much more personal and complex. Only in recent history have individual codes, passwords, usernames, etc., become necessary. Every individual with a bank account or credit card, or who uses a cellular telephone or a computer, is under attack every day by anonymous digital criminals capable of stealing and exploiting all the individual's assets and personal information for financial gain or even worse. It is no longer only a military or political objective to create and decipher codes. Digital thieves hold or protect no soil, nor do they fly a flag. They have no honor. They are nothing more than common thieves. They hide in digital shadows and prey on innocent and largely defenseless people of all types just to steal their money or otherwise cause them harm.

Angel Communication Code explores a new and different kind of cryptology. It is about the discovery of a code that is for the benefit of all humanity, one that is based not only on the patterns found in mathematics and biology, but also in the messages of scripture. This new decryption requires an examination and identification of a different kind of messenger with a different purpose and a different technique. There is an entity out there that wants us to find the message but that will not simply hand over the keys. This entity requires us to demonstrate that we can figure it out and repeat its message back to it consistent with the concept of three-way communication:

1. Author sends the message.
2. Receiver finds the message and repeats it back to the author.
3. Author confirms to the receiver that the receiver has understood the author's message.

Humans are on step 2. We must find the message and repeat it back to the originator(s). After that, the communication barrier will be broken and lines of communication will begin to open up.

People do not create, send, or seek coded messages without a reason. Depending on what that reason might be, the message seeker will be directed to a particular place to discover a coded message. We are seeking a different sort of message provided by or through angels and ETs that supports a hypothesis about ET communication. This requires us to take a closer look at angels and their purpose.

[31] Patrich Chisholm, "Computer Code in Nature Points to an Ultimate Programmer," Seventimes70, December 19, 2018.

CHAPTER 3

What Are Angels?

This is not a simple question, and you might be surprised at the variety of answers that are returned when one asks this directly. Angels are a fascinating topic in scripture and a complex subject of debate in the study of theology. Angels play an important role in Christian scripture, from the first verses of the first book of the Old Testament (Genesis) to the closing verses of the last book of the New Testament (Revelation). Angels have made a significant impact on humanity throughout the history of God's people as told in the Bible, but what are they exactly, what is their purpose, and how did they come into existence?

Going back to basics to get some of these answers will require that we revisit and rethink the contemporary understanding of angels. Trying to define and understand what an angel is turns out to be an extraordinarily complex and sensitive business, and not all theologians agree upon certain aspects of this topic. The question about when the creation of angels happened is an unsolved mystery and the subject of endless debate in theological circles, as the Bible is not forthcoming or clear on this point. Angels are referenced nearly three hundred times in the Bible, and given that such is the case, their existence and their individual messages are obviously important to the overall Bible message and the Christian faith. Angels are a most fascinating topic and are referenced throughout scripture for a reason. They were created to do God's will and deliver His messages. A message is what we seek in *Angel Communication Code* as it applies to two-way communication with our ET brothers and sisters.

In the very first book of the Bible, Genesis, scripture tells us that angels were instituted as guards of the gates Eden after Adam and Eve were banished by God for eating of the tree of knowledge (of good and evil). The archangel Gabriel announced to Mary that she would be the mother of Jesus. She was chosen by God to be the mother of His only Son, the Christ, and God chose an angel to deliver that message rather than delivering the message directly to Mary Himself. This seems a bit indifferent of Him given the magnitude of the required assignment placed upon Mary. It is not as though Mary had a choice in the matter. God did deliver messages directly to many others, such as Moses and Noah. In fact, God the Father never speaks directly to a woman in the Bible, with the exception of Eve, the first woman of the scriptures. Even the message of Mary's impregnation and the gestation and birth of the Savior of humanity, His only begotten child, He delegated to hands of an angel. That fact alone should tell us something about the importance of angels. At the risk of being blasphemous, I'll venture to say that it also tells us something about God's deeper understanding of human intimacy and emotion or His apparent lack thereof (at times).

The sources of information we have about angels starts with the Old Testament, then moves through the New Testament and, in scholarly circles, extends to Second Temple–period literature, which dates from the fifth century BC to the first century AD. Perhaps the most influential and extraordinary of the extrabiblical writings is the book of Enoch, which is quoted in the book of Jude. Even though scripture does not tell us exactly when angels were created, it does give us information about their nature and a few hints as to their existence before humanity was created.

We know that angels are spirit beings (Hebrews 1:14) and that they are capable of both visible and invisible forms as cited in Numbers 22:22–35. We know that they are capable of emotion (Luke 15:1–10) and intelligence (2 Samuel 14:20), and that they are powerful (2 Peter 2:11) and immortal (Luke 20:34–36).

Scripture provides us with only glimpses with respect to their duties:

- Angels worship the Lord (Revelation 4:8, 5:11–12).
- Angels are ministers to believers who die in Christ (Jude 9; Luke 16:22).
- Angels rule nations (Daniel 10).
- Angels care for believers in times of difficulty and trial (Matthew 4:11).
- Angels strengthen us to proclaim the gospel and stand strong in faith (Luke 22:43).
- Angels bring God's judgment to both individuals and entire nations (Acts 12:23; Genesis 19:3; 2 Samuel 24:16; Revelation 16:1).
- Angels intercede on our behalf but only as directed by God (Zechariah 1:12; Revelation 8:3–4).

Scripture does tell us that angels existed before humankind was created. Genesis 3 speaks of a demon angel known only as "Nachash," which in Hebrew translates as "serpent," though it may be more accurate to translate it as "shining one." Nachash lies to Eve, convincing her that she will become like God by eating the fruit the Lord had forbidden. Other scripture verses confirm the prehuman existence of angels.

Dr. Mike Heiser, author, biblical scholar, and executive director of the Awakening School of Theology and Ministry, writes that at least some of the angels were a group known as the divine council. He explains, "The term 'Divine Council' is used by Hebrew and Semitic scholars to refer to the heavenly host (army), the pantheon of divine beings who administer the affairs of the cosmos." He references Psalm 82:1 as just one place in the Bible where there is mention of this divine council.

Heiser's research is controversial and not without its critics, but it is food for thought without a doubt. In the end, while we do not know the exact moment of angel creation, we do know that angels have existed longer than humankind and that at the creation of human beings, they became humankind's spiritual benefactors.[1]

Many people believe that when a saved person goes to heaven, he or she is given an opportunity to become an angel; however, there is no biblical basis for such a belief. Angels and humans are two unique and separate entities. The Bible says that humankind is slightly lower in holy stature than are angels (Hebrews 2:5–7). The Bible also teaches that through redemption and salvation, humankind becomes superior to angels and that the saved will one day judge the angels.

Lucifer was first created as one of God's most glorious angels, "full of wisdom and perfect in beauty" (Ezekiel 28:12). Lucifer's heart, however, became swollen because of his beauty (Ezekiel 28:17), and his ego caused him to try to seize the throne of God (Isaiah 14:12–14). This got him cast out of heaven by God, and one-third of the angels were cast out with him because they joined his rebellion (Revelation 12:4). Lucifer

[1] Jack Ashcraft, "Do We Know When Angels Were Created?," Christianity.com, March 12, 2021.

came to be called Satan after that happened. A literal throne is a seat of power and the highest place of authority in any given realm. It's a place from which a ruler reigns, acts on behalf of his or her people, and exercises justice. Unlike earthly monarchs, who are bound by time and territory, God's throne represents His transcendent reign over the whole universe for all eternity. Because God exists in spirit form, He has no need to physically sit upon an ornate ceremonial chair. The Bible gives detailed information about the throne of God as the literal and figurative symbol of His all-consuming glory and infinite sovereignty.[2]

The Bible clearly indicates that angels can take on a human image and appear before people. In the Old Testament there are recorded appearances of angels to Abraham, Hagar, Lot, Jacob, Moses, Joshua, Gideon, David, Daniel, and many others. The New Testament opens with a whole series of angelic appearances related to the birth of the Messiah. After the death of Jesus, angels appeared in His tomb and at His ascension. Peter, John, Philip, and Paul all had angelic encounters in their ministries.

When performing as God's messengers, angels can manifest themselves to humans as humanlike males. The notion of angels having wings is limited to the Bible's description of two special angelic beings called seraphim and cherubim. These angels are said to dwell in the throne room of God in heaven. Seraphim are described as having six wings (Isaiah 6:2). Cherubim are depicted with four wings (Ezekiel 1:5–6).

When angels take on human form, they appear like any normal person (Genesis 18:2, 19:1–17). When they manifest themselves in their spirit form, they tend to appear as a spectacular light (Matthew 28:1–4).

Angels have emotions as demonstrated by their rejoicing over the creation of the universe (Job 38:7). The Bible also tells us that "there is joy in the presence of the angels of God over one sinner who repents" (Luke 15:10).

Angels do not marry or reproduce (Matthew 22:30). They do not get old, and they are not subject to death (Luke 20:36). This implies that their population remains constant. The exact number of angels is never specifically quantified. Ten thousand angels appeared to Moses on Mount Sinai (Deuteronomy 33:2). David saw twenty thousand at one time (Psalm 68:17). When John experienced the throne room of God, he estimated that he saw ten times ten thousand angels, to make a point about the magnitude of their numbers.

Angels are said to possess great knowledge, but they do not possess infinite wisdom, understanding, and insight. Angels are said to be very powerful (2 Thessalonians 1:7; 2 Peter 2:11), but they do not have unlimited power to do anything they choose. They can move about rapidly, but they are not everywhere at all times. Neither are they omniscient, nor are they omnipotent or omnipresent. They are obedient servants of their Creator. Angels are not gods and are not to be worshipped (Colossians 2:18; Revelation 22:8–9).

There is no biblical reference claiming that angels have to eat to stay alive, but the Bible does say that they do eat when they take on human form (Genesis 18:1–8, 19:1–3).

When angels speak to people, they always use that person's native language, as we would expect. The Bible also states that angels have a language of their own that they use among themselves (1 Corinthians 13:1).

---

2 Annette Griffin, "What Is the Throne of God and Does It Really Exist?," *Christianity Today*, May 17, 2021.

Regarding the host (army) of regular angels, the Bible makes it clear that they are servants of God who have a variety of tasks. The very word *angel* is derived from the Greek word *angelos*, which means "messenger." The Hebrew equivalent word, *malakh*, also means "messenger." Angels are portrayed in the scriptures as delivering messages from God. Examples of this are numerous and woven throughout the scriptures. A few critical examples are as follows:

- The laws (commandments) were given to Moses via angels (Acts 7:53; Galatians 3:19).
- Angels are presented as mediums of revelation to God's prophets (Daniel 4:13–17; Zechariah 1:9–11; Hebrews 2:2).
- Abraham was informed through angels that his barren wife would have a child (Genesis 18:1–10).
- Lot was warned by angels of the impending destruction of Sodom and Gomorrah (Genesis 19:1–16).
- Joseph was visited by angels three times in dreams. He was informed that Mary was in fact a virgin and was carrying the Son of God (Matthew 1:20). He was also told by an angel to flee with his family (which included Jesus) to Egypt (Matthew 2:13). In addition, he was informed via an angel when it was safe to return home to Nazareth (Matthew 2:19–20).
- The very first Gentile convert, Cornelius the Roman centurion, was directed by an angel to send for Peter to receive a special message from God (Acts 10:1–8).

Another significant task of angels is to minister to the needs of saints. They were sent to provide food and drink for Elijah as he lay dying in the wilderness (1 Kings 19:1–8). They were also sent to tend to Jesus after His forty days of reflection and temptation in the wilderness (Matthew 4:11). An army (host) of angels surrounded the prophet Elisha to protect him when his life was endangered (2 Kings 6:15–19). In the New Testament, we are told that Peter was released from prison by an angel (Acts 12:5–11).

In addition, and equally as significant, angels are charged with the task of executing the judgments of God. They were sent to destroy Sodom and Gomorrah (Genesis 19:12–15). It was an angel that killed 185,000 Assyrian soldiers in one night to prevent them from attacking Jerusalem (2 Kings 19:35). Psalm 78:49 speaks of a "band of destroying angels" that was sent to torment the Jews in the wilderness when they rebelled against God. The New Testament makes it clear that God will execute His end-time judgments on the nations through angels (Matthew 13:49–50; Jude 14–15).

The Bible reveals that there will be a flurry of angelic activity in the end times. The four living creatures before the throne of God are the ones who will execute the first series of judgments that will mark the beginning of the Tribulation. The Tribulation is a future seven-year period of time when God will finish His discipline of Israel and finalize His judgment of the unbelieving world. The Christian church, made up of all who have trusted in the work of the Lord Jesus to save them from being punished for sin, will not be present during the Tribulation. The Christian church will be removed from the earth in an event cited in the Bible as the Rapture.

It is in response to the commands of angels that the four horsemen go forth with the seal judgments that result in the death of one-fourth of humanity (Revelation 6:1–8). At the same time, it is angels who will protect 144,000 Jews supernaturally (Revelation 7:2–3). When the second series of judgments begins (the trumpet judgments), it will be angels who serve as the instruments of God's wrath, resulting in the deaths of one-third of those still left alive (Revelation 8–9).

In the middle of the Tribulation, Satan will try one last time to seize the throne of God, resulting in a war in the heavens between Satan with his demons versus the archangel Michael and his holy angels. Satan will be defeated. He and his demons will again be cast down to earth. His access to heaven will be cut off permanently (Revelation 12:7–12).

In the second half of the Tribulation, we are told that God will give humankind one last opportunity to repent before He pours out His final wrath. He does this in a most peculiar manner. He sends three angels with very specific tasks. The first angel is given the responsibility to proclaim the "eternal gospel" to "every nation and tribe and tongue and people" (Revelation 14:6). This will be the only time in history that God allows an angel to preach the gospel and tasks that angel with preaching the gospel. The second angel will follow the first as he travels the earth with the message. He will proclaim the impending destruction of the kingdom of the Antichrist (Satan) (Revelation 14:8). The third angel will follow along behind the first two. His role will be to issue a warning that anyone who worships the Antichrist and receives his mark (666) will suffer God's wrath in hell for eternity (Revelation 14:9–11). These three proclamations by three angels are followed by the final outpouring of God's wrath, called the bowl judgments (Revelation 16). Once again, seven angels will execute these judgments (Revelation 16:1). Angels are also the ones who will destroy the capital city of the Antichrist in one hour on one day (Revelation 18).

In all of this, it is important to recognize that the Bible is not the verbatim Word of God as commonly proclaimed. The Bible was written by approximately forty men with a great diversity of backgrounds over the course of approximately fifteen hundred years. Not all the books of the Bible specifically identify the author. The differences within the Bible with respect to the portrayal of angels and the way humans depict angels are a clear clue that perhaps the authors of the Bible and their accounts of angels within the Bible were opportunities to insert other meanings by way of hidden messages from angels. The message we can extract is the frequent occurrence of prime numbers 2, 3, 5, and 7.

The overall message of the Bible has remained amazingly consistent over the centuries, but there have been changes to the Bible with respect to which books are included or excluded within each version. There are approximately forty-five thousand worldwide denominations of Christianity, each with its own twists. In all the biblical references to angels, there is not one single specific reference about angels in any Bible version that is about the angels in and of themselves. Angel references in the Bible are always incidental to some other situation or impending event. The Bible never deliberately speaks about or attempts to explain the nature of angels. They are just there. Angels are deployed by God only to inform us of God's message, what He is doing at the time of their inclusion in scripture, what He does in the grand scheme of the universe, and how He does or will do those things. Angels only exist to deliver messages from God.[3] This makes understanding the origin of angels an extraordinarily complex subject to address at any level, from church preaching on Sunday to the intensive study of theology.

Throughout history, angels have often been depicted in artworks as both male and female. When angels are referenced in the Bible, however, they are always spoken of in the masculine form. In addition, when angels are said to have appeared to people in the Bible, they were always described as male. The Hebrew word *malach* and the Greek word *angelos* are both masculine nouns. The only specifically named "good" angels in the Bible are male: Gabriel and Michael. There are two "fallen angels" specifically named in the

---

3 Millard J. Erickson, *Christian Theology* (Grand Rapids, MI: Baker Book House, 1983), p. 434.

Bible, which are Lucifer (Satan) and Abaddon (Apollyon)—also male. Abaddon is mentioned only one time, in Revelation. The name Apollyon means "destruction."[4]

There are some theologians and church clergy who interpret Zechariah 5:5–11 and the two women referenced with wings as being female angels.

> Then the angel who was speaking to me came forward and said to me, "Look up and see what is appearing." I asked, "What is it?" He replied, "It is a basket." And he added, "This is the iniquity of the people throughout the land." Then the cover of lead was raised, and there in the basket sat a woman! He said, "This is wickedness," and he pushed her back into the basket and pushed its lead cover down on it. Then I looked up—and there before me were two women, with the wind in their wings! They had wings like those of a stork, and they lifted up the basket between heaven and earth. "Where are they taking the basket?" I asked the angel who was speaking to me. He replied, "To the country of Babylonia to build a house for it. When the house is ready, the basket will be set there in its place." (Zechariah 5:5–11)

The confusion is that in the prophecy, two women with wings fly away with the basket of wickedness. The women are described with wings of a stork (considered an unclean bird) but are not specifically referred to as angels. Zechariah's prophecy is laden with metaphors and imagery, not intended to be taken literally. The prophecy, cloaked with metaphors, is painting the picture that "wickedness was borne swiftly from the land."

All the authors of the Bible books are believed to be male. The Bible never states that female angels exist or that angels have gender. In the days of each Bible author, the male gender had the leadership role in the home, the society, and the church. It is most likely that angels were given male names by the Bible authors to indicate that they also have authority as God's representatives. This again speaks to how recorded history tells us as much about the author as it does about the subject. That, however, cannot be the only rationale behind all the places in the Bible referring to angels in masculine form and with masculine names.

In art and culture, female angel depictions in Christian works can be attributed to female depictions of angel-like entities from ancient paganism that could have woven their way into Christian thinking and Christian art. Pagan religions often included servants of the gods that were clearly female. They are not angels but, rather, pagan goddesses that have wings and are said to act in a similar way as Christian angels, that is, making appearances and delivering messages. Pagans worshipped idols with wings. The Greek female winged goddess Nike, for example, was considered the messenger of victory.

It is reasonable to assume that angels, however they appear in Christian scripture, are not male or female in any sort of human context as humans understand and experience gender for reproductive reasons. Yet Christian cultures have historically depicted some angels clearly as female in nonbiblical writings and in the visual arts. The reality is that the Bible identifies angels exclusively in masculine terms.[5]

There are no biblical references about angels reproducing to make baby angels, nor are there any biblical references to the creation of new angels in any way. The Bible says nothing about people going to heaven after they die to earn their wings and then become angels. This perception makes for an interesting

---

[4] Amy Swanson, "What Are All the Names of Angels in the Bible?," Christianity.com, October 28, 2020.

[5] Emily Hall, "Are Angels Male or Female?," Christianity.com, March 12, 2019.

bedtime story or motion picture theme, but it is not accurate. Angels are spiritual beings created by God for God's service. They are different creatures from humans. In Hebrews 1:14 it is said: "Are not all angels ministering spirits sent to serve those who will inherit salvation?"[6] It is generally accepted by theologians that the universe started with one God who created a finite number of angels. Angels are immortal and new angels are not created according to scripture.

There are references in the Bible to fallen angels coming to earth and mating with human women who bore their children. Their children were called the Nephilim, a race of so-called giants. This sort of reference comes early in the first book of the Bible:

> And it came to pass, when men began to multiply on the face of the earth, and daughters were born unto them, that the sons of God saw the daughters of men that they were fair; and they took them wives of all which they chose. And the Lord said, My spirit shall not always strive with man, for that he also is flesh: yet his days shall be an hundred and twenty years. There were giants in the earth in those days; and also after that, when the sons of God came in unto the daughters of men, and they bore children to them, the same became mighty men which were of old, men of renown. And God saw that the wickedness of man was great in the earth, and that every imagination of the thoughts of his heart was only evil continually. And it repented the Lord that he had made man on the earth, and it grieved him at his heart. And the Lord said, I will destroy man whom I have created from the face of the earth; both man, and beast, and the creeping thing, and the fowls of the air; for it repenteth me that I have made them. But Noah found grace in the eyes of the Lord. (Genesis 6:1–8)

The first book of the Bible describes the creation of heaven and earth and of humankind. It was only six chapters into the first book of the Bible when God said he would destroy it all because some of the sons of God (angels) fell, became physical (not spiritual) beings, and mated with the women of His human creation, resulting in a race of wicked people. All of this suggests that good angels always remain spiritual beings and, as such, cannot impregnate human women; however, fallen angels lose their spiritual form, are able to take on biological humanlike form, and can impregnate women.

It is this concept that makes the Immaculate Conception of Jesus as the only begotten Son of God and his human mother, Mary, so incredibly important to the Christian faith. There is great irony in the fact that fallen angels took on a form that would allow them to breed with human women and the result was something God chose to destroy. Conversely, God without human form, somehow with His great power, impregnated a human woman, and the result was Jesus, who is the Savior of humankind and the path to eternal life in heaven. Jesus, the part-human Son of God, is the King for eternity. The part–fallen angel, part-human creatures called the Nephilim were destroyed.

As cited previously, the Bible says that a third of all angels became fallen angels. Stating that one-third of the angels fell implies that there is indeed a finite number of angels. There is no such thing as a third of an infinite number, at least as humans currently understand mathematics. The prevailing theological interpretation is that this accounting comes from Revelation 12:3–4, which states:

> And there appeared another wonder in heaven; and behold a great red dragon, having seven heads and ten horns, and seven crowns upon his heads. And his tail drew the third

[6] Jason Soroski, "What Do We Know about Angels in the Bible?," Bible Study Tools, April 23, 2020.

> part of the stars of heaven and did cast them to the earth: and the dragon stood before the woman, which was ready to be delivered, for to devour her child as soon as it was born.

The majority of theologians interpret this verse as Satan (red dragon with seven heads) bringing down the fallen angels equal to one-third of all the angels (one-third part of the stars). Theologists interpret the "one-third of the stars" in Revelation 12:3–4 in a variety of ways. The Bible, therefore, may or may not support the idea that one-third of the angels fell with and because of Satan, depending on precisely what John meant in composing Revelation 12:4.[7]

The question that needs to be asked is, about from where did the 1:3 ratio derive, and what is its significance to the subject of *Angel Communication Code*? It is not 1:2 or 1:5; it is precisely 1:3, and not approximately 1:3.

Recall again that the primary driver of the extraterrestrial communication code is the number 3. The ratio of 1:3 is 0.33333333 with infinite decimal places of the number 3. This would mean that two-thirds of good angels must therefore remain. The ratio 2:3 is 0.6666666 with infinite decimal places of the number 6. Perhaps the infinite decimal places represent the infinite universe of God's creation. This is very curious in that 3 is the number driving our code and 666 is the number physically marking Satan, the beast in the book of Revelation. Infinite 3's are attached to fallen angels, and infinite 6's are attached to both the king of fallen angels (Satan) and the balance of the good angel population. That is ironic, weird, and a little frightening. Also note that expressing the first and third prime numbers as ratios is 1:2 and 1:5: 1/2 = 0.5, and 1/5 = 0.2, with only one decimal place for each ratio, the first and the third prime number (2, 5). The second prime number, 3, as a ratio of 1:3, is completely different.

Let us organize and summarize these numerical relationships and perhaps see why they matter:

> 1/2 = 0.5—the first prime as denominator = the third prime decimal to one decimal place that is the third prime number
>
> 1/3 = 0.333333—the second prime as a denominator = the second prime as decimal to infinite decimal places of the second prime number
>
> 1/5 = 0.2—the third prime as denominator= the first prime decimal to one decimal place of the second prime number; the 2, 3, 5 loop is closed
>
> 2/3 = 0.666666—the 2:3 balance of good angels versus bad angels = Satan's mark to infinite decimal places
>
> 3/5 = 0.6—the second prime over the third prime = the first digit of Satan's mark to one decimal place
>
> 5/3 = 1.666666 (or 1.667)—the third prime over the second prime number = Satan's mark to infinite decimal places. It is also very close to the divine ratio (1:62) previously discussed.

[7] Matt Slick, "Where in the Bible Does It Say That One-Third of the Angels Fell?," Christian Apologetics and Research Ministry, August 9, 2009.

While all of this is being extracted from the last book of the Bible (Revelation), again consider that Genesis (book 1 of the Bible) 6:6–7 tells us: "The Lord regretted that he had made human beings on the earth, and his heart was deeply troubled. So the Lord said, 'I will wipe from the face of the earth the human race I have created—and with them the animals, the birds, and the creatures that move along the ground—for I regret that I have made them.'"

This in essence closes the Bible loop. The prime number code loop and the Bible beginning and end loop are closed. It is the Alpha and Omega concept. Putting any of the first three prime numbers in play as a numerator also brings Satan into play. It also detaches our analysis from a prime number analysis and points us toward the first book of the Bible along with the last book of the Bible, creating and closing that loop.

If someone wanted to drop a heavy clue that focuses a person's attention on prime numbers 2, 3, 5, and 7, then Revelation, which describes the end of the world, would be a good choice of location. It is curious, however, that the 2:3 ratio part of that description, the population balance of good angels, is left for us to deduce without any specific direction from the Bible author of that story. This whole examination is a very curious and clever biblical dance between angels, demons, and the first thee prime numbers that are our code drivers. It is not a coincidence. It must surely have been put into play by intelligent design. The numbers cited in the book of Revelation lead us into the discussion about Armageddon as described in Revelation.

Armageddon is a place, and not an event as is often thought. The word *Armageddon* is derived from the Hebrew word *Har Megiddo*, which means "the mountain of Megiddo." Over many centuries, multiple nationalities and languages added an *n* and dropped the *h*, transforming *Har Megiddo* to *Harmageddon*, which morphed into the word *Armageddon* over time.

Megiddo was inhabited approximately between 7000 BC and 300 BC. Many battles were fought near Megiddo during that time span. The book of Revelation, which refers to the site as Armageddon, prophesies that a final battle at the end of time would take place there.[8]

The final battle at Armageddon is not the end of time. It is the place where the final battle between good and evil takes place. Good is prophesied to win, and then Jesus will reign over the world for a thousand years. The thousand-year reign of Jesus, also called the millennium or the millennial kingdom, is the period between the Tribulation and the destruction of the world. This does not mean that the earth dissolves into nonexistence. It means the end of the world as we know it. The purpose of the thousand-year reign is to allow for the fulfillment of several promises.

God made several unilateral promises, or covenants, promising to bless Israel in specific ways. Some of these covenants have yet to be fulfilled. The Palestinian Covenant (Genesis 15:18–20; Numbers 34:1–12) defines the exact borders Israel shall possess for eternity. The nation of Israel has yet to encompass these specific boundaries. The Davidic covenant (2 Samuel 7) is the promise that God made to David that David's heir would sit on the throne of Israel forever. Jesus is that heir, but He has yet to take His physical, political birthright in that capacity. The new covenant (Jeremiah 31:31–34) outlines the promise that Israel as a nation will return to God and worship their Messiah (Romans 9–11). These covenants will be fulfilled during the thousand-year reign.[9] It is all very confusing. The world is prophesied to change dramatically,

[8] Owen Jarus, "Welcome to Armageddon: Meet the City behind the Biblical Story," Live Science, May 5, 2020.

[9] "The Truth about the End," Compelling Truth, https://www.compellingtruth.org/1000-year-reign-Christ.html.

but is it actually going to end at some point or not? The question we ask is, how and when will the time of humans end?

What makes this relevant to *Angel Communication Code* is that clearly something is about to happen between humans on earth and beings not of this earth. We are seeking the path to two-way communication with ET beings from other planets and galaxies, and clues to that path could be buried in the Bible and being revealed to us via angels—God's messengers.

Theologians mostly agree that John the apostle (Saint John) is the author of the book of Revelation; however, there is small faction of theologians who have challenged this notion without much credibility given to their challenge. John is regarded by the Christian church as the author of five books of the Bible:

- the Gospel of John
- 1 John
- 2 John
- 3 John
- the book of Revelation

He is also believed to be the only disciple who died of old age. All the others were allegedly martyred. Ancient sources may or may not refer to the apostle John by several other names, including John of Patmos (because he was banished to the island of Patmos), John the Evangelist, John the Elder, John the Presbyter, and the Beloved Disciple. It is unclear if all (or any of) these names do in fact refer to Saint John. We do know that John the disciple of Jesus is not the same person as John the Baptist, who was believed to be Jesus's cousin.[10] John the disciple was one of the first and closest disciples of Jesus. He was a very important person in the time of Jesus and after the crucifixion of Jesus. His words carry a lot of weight, and his words in Revelation have led us to a strong reference to the numbers 3 and 7 regarding angels. He also has a significant link to the number 5.

If the primary purpose of angels is to worship God, serve God, and be messengers for delivering God's messages and instructions to humans, then it makes a lot of sense that someone who wanted to leave future generations a message about establishing contact with a different flock of God's creation who dwell on another planet or in another universe—or even in another realm/dimension—would use angels to deliver that message. What more credible source for a message could there possibly be? The opportunities for the inclusion of an additional message with respect to the subject of *Angel Communication Code* lie in the following:

a. the conspicuous lack of angel details in the Bible itself and from modern theologians on the overall subject of angels in general;
b. the method by which the message is being delivered (there is God's message to humans delivered through angels, and then there are the clues implying numerical keys to an encrypted message buried somewhere in the history of angels).

The method of angel message delivery we are alluding to in *Angel Communication Code* is not what is typically cited in the Bible. The message is not so obvious. The entire concept of *Angel Communication Code*

10 Ryan Nelson, "Who Was John the Apostle?" OverviewBible.com, February 6, 2019.

is about figuring out in what form the message was delivered and how to decipher that message and then use it for its intended purpose.

More than two thousand years after the mortal death and subsequent divine resurrection of Jesus, people cling to the belief in angels. A report by *CBS News* found that the belief in angels is primarily tethered to religion, as one might expect, but not exclusively so. The belief is actually held by a much broader audience. CBS researchers found that 88 percent of Christians, 95 percent of evangelical Christians, and 94 percent of those who attend weekly religious services of any sort say that they believe in angels. The belief in angels is also fairly widespread even among less religious people. A majority of non-Christians believe angels exist, as do more than 40 percent of nonbelievers of any faith who never attend religious services.[11] Angels or their equivalent are not only found in Christianity. There are angels in the Islamic faith who serve Allah in the same way Christian angels serve God as messengers. Hinduism includes a variety of spiritual beings who act in angelic ways. Buddhism has devas, which are celestial beings that perform the messenger function for Buddha. The overwhelming majority of the world's population, approximately 80–85 percent, believes in angels in some form.

A 2011 poll conducted by the Baylor University Institute for Studies of Religion has found that more than half of all Americans believe they have been directly helped specifically by a guardian angel in the course of their lives. Of the seventeen hundred respondents, 55 percent answered affirmatively to the statement, "I was protected from harm by a guardian angel." The responses did not delineate between standard class and denominational assumptions about religious belief. The majority held up regardless of denomination, region, or education.[12] The problem with this poll is that it was specifically about guardian angels and not just angels. The idea of divine guardian angels is a much larger and more opaque subject than one might expect.

The idea of personal guardian angels is a broadly held belief among Christians and others. This belief, however, is not a belief necessary for salvation and is not binding upon Christians. In fact, the number of doctrinal subjects that the church has defined as definitively true in general is relatively low. Regarding angels, there are not many things that the church can say it definitely knows without question. The more important things that they say they do know are as follows:

a. Angels are not gods; they are spiritual creatures.
b. Angels were originally created to be good by God.
c. Angels have the power of an intellect and a will, and when some of them used their will to defect from God, they became fallen angels, falling from the goodness with which they were created.
d. Angels are engaged in a great cosmic war: the good and heavenly angels versus the fallen angels, also known as demons.

Those are the very basics about angels.

Guardian angels are not explicitly defined by the church or and don't explicitly appear in scripture; however, there are passages in scripture that suggest they might exist. In the Old Testament, there are numerous examples of angels of the Lord sent to guard and protect humanity. Scripture, however, never cites that individuals have a specific angel assigned to them as a personal guardian angel. The author of Psalm 91

---

[11] CBS News, "Nearly 8 in 10 Americans Believe in Angels," December 22, 2011.

[12] David Van Biema, "Guardian Angels Are Here, Say Most Americans," *Time*, September 18, 2008.

proclaims: "For he commands his angels with regard to you, to guard you wherever you go. With their hands they shall support you, lest you strike your foot against a stone." This statement, however, is not referencing or identifying a unique individual guardian assigned to one person.

From the book of Tobit it could argued that the angel Raphael acted in the capacity of a guardian angel. Tobit, also called the book of Tobias, is an apocryphal work (noncanonical for Jews and Protestants) that found its way into the Roman Catholic canon via the Septuagint (the Greek Old Testament). A religious folktale and a Judaized version of the story of the grateful dead, it relates how Tobit, a pious Jew exiled to Nineveh in Assyria, observed the precepts of Hebrew law by giving alms and by burying the dead. In spite of his good works, Tobit was struck blind.

Concurrent with Tobit's story is that of Sarah, daughter of Tobit's closest relative, whose seven successive husbands were each killed by a demon on their wedding night. When Tobit and Sarah pray to God for deliverance, God sends the angel Raphael to intervene and protect. Tobit regains his sight, and Sarah marries Tobit's son Tobias. The story closes with Tobit's song of thanksgiving and an account of his death.[13] Many argue this is the go-to reference for proof of the existence of guardian angels.

This story of Raphael is not an exact parallel to guardian angels as they are understood today; however, it is the one case where an angel was definitely set on a very particular mission to protect one person. The opponents of the existence of guardian angels will say it was one mission for one guy. Raphael did not stick around after the mission was completed to be Tobit's personal guardian.

In the New Testament, there is also arguable, albeit circumstantial, evidence for the existence of guardian angels:

- In Matthew 18:10, Jesus tells His disciples: "See that you do not despise one of these little ones, for I say to you that their angels in heaven always look upon the face of My heavenly Father."
- In Acts 12:15, Peter, just freed by an angel from prison, visits Mary's (the mother of Mark) house. While Mary is convinced she sees Peter, her houseguests tell her that she must be mistaken and instead might be seeing his angel.

The key that makes both these cases interesting when thinking about guardian angels is the possessive form. In Matthew 18, it is "their" angels, and in Acts, "his angel" is spoken of.

Acts 12 also demonstrates that the earliest Christians had a belief in personal angels assigned to guard and protect specific people. The idea of, and belief in, guardian angels has been around for a very long time, but angels are not specifically identified in scripture. If there were such a thing as a guardian angel, we would not have to search so hard for a biblical reference to them. Their existence would be made clear and obvious.

Putting scripture aside, early church fathers addressed guardian angels in their writings and homilies. Origen of Alexandria, a third-century theologian, was perhaps one of the most prolific writers on the subject of guardian angels. It was Origen who talked about two angels that "attend each human being" in his homilies on Luke. Origen wrote in one of his homilies on Luke: "One is an angel of justice, the other an

---

[13] *Encyclopedia Britannica*, s.v. "Tobit," December 27, 2019, https://www.britannica.com/topic/Tobit-biblical-literature, accessed May 22, 2022.

angel of iniquity. If good thoughts are present in our hearts and justice springs up in our souls, the angel of the Lord is undoubtedly speaking to us. But, if evil thoughts turn over in our hearts, the devil's angel is speaking to us."

The church has not officially agreed that the so to speak "left shoulder / right shoulder angel" theory is correct, but it does confirm that the early church fathers did not deny the concept of a personal angel. Origen was a guy who believed in both personal guardian angels and also therefore guardian demons fighting to control a person's fate. It is the manifestation of the struggle between good and evil in all of us. It is said that Origen derived much of this thinking from the Shepherd of Hermas. This is an important early Christian text that some church fathers considered to be canonical, though it has never been officially included in the Roman Catholic canon.

The Shepherd of Hermas is a story about a liberated slave named Hermas. He was visited by an angel while he was praying at home. Hermas said that a man of glorious aspect dressed like a shepherd walked into the room, saluted him, and said, "I've been sent by a most venerable angel to dwell with you the remaining days of your life." So, we get this idea of guardian angels in some of the earliest postbiblical literature in the second century. This would have surely influenced the thinking of Origen.

Saint Jerome, a fourth-century church father, theologian, and doctor of the church, famously said of guardian angels: "How great the dignity of the soul, since each one has from his birth an angel commissioned to guard it."

The catechism passage on guardian angels talks about an angel standing beside each believer, which lends credence to the theory that angels are assigned to baptized individuals only. The tradition of the church leans toward the idea that all people, regardless of baptism status, have a guardian angel. This is stated explicitly by Saint Thomas Aquinas in the Summa Theologica. Saint Thomas Aquinas also favored the understanding that guardian angels appear to individuals at their birth. This idea is not stated in scripture, however.

Whatever the sidebar arguments may be, the Bible makes no obvious, direct reference to guardian angels. The Bible never specifically names or otherwise calls out a guardian angel that is specific to any one person. God assigns an angel to assist and protect individuals as one of an angel's many missions. It is quite comforting to believe that you have your own personal guardian angel dedicated to you, which is probably why most people tend to believe in them. It is an immensely popular and comforting concept in today's world and historically. Entertainment outlets have promoted the idea of guardian angels and have emphasized the concept that each person has their own guardian angel. Going through life with the idea that you have your own guardian angel is extremely common. Stories from people about their guardian angel experiences are common in all parts of the world from Christians, people of other religions, and nonbelievers alike. Although a belief in guardian angels is quite common, God, not an angel, is the one who protects us. In times of need, and when God chooses to lend His support, He sends an angel to execute His bidding. This of course is not always the case. God has also directly intervened for a select few, and not through an angel. Angels do not step in on their own whenever they feel it is necessary or upon a person's request. That is God's decision and His alone.

Perhaps it is possible that what is being perceived as a guardian angel is a deceased loved one who has the ability to look over one's shoulder here on earth from another realm.

The Roman Catholic Church celebrates the feast of the holy archangels Michael, Gabriel, and Raphael every year on September 29. Note here again that Michael and Gabriel are the only two angels mentioned by name in the Bible. Raphael is not. Archangels are the second-lowest rank in the third (lowest) angel order. Archangels are called such because their choir is just above that of the lowest angels, simply called "angels." Both archangels and angels are the choirs charged with the task of delivering God's messages to humans. The archangels are seen to be the highest of all God's messenger angels to humans, and they are given the most important missions. It is said that there are only seven archangels and that Michael is regarded as the highest ranking. In the book of Tobit (not a book of the Bible), when Raphael reveals his identity, the archangel says, "I am Raphael, one of the seven angels who stand and serve before the glory of the Lord" (Tobit 12:15). The passage from Tobit then begs the question, if Michael, Gabriel, and Raphael are three of the archangels, what are the names of the other four? This question has been taken up by various branches of Christianity, who often look to the book of Enoch for this answer. The book of Enoch is an ancient Jewish text that is not accepted as canonical in either Hebrew scriptures or the Christian Bible. It is written in the Bible that Enoch is one of two people never to have died. The other was Elijah. Enoch was an Ethiopian Jewish man. These two men never were said to have died but, rather, are explicitly said to have been taken alive by God to walk with Him.[14] In those days, walking with God did not refer to dying and going to heaven. When someone died, they were said to have "given up the ghost." Seeing as Enoch's book was written by a man who is said to have "walked with God," it seems odd that it was not included in Christian Bible sometime around the fourth century AD. The book is not part of the biblical canon used by Jews, with the exception of Beta Israel (Ethiopian Jews). While the Ethiopian Orthodox Tewahedo Church and the Eritrean Orthodox Tewahedo Church consider the book of Enoch as canonical, other Christian groups regard it as noncanonical or noninspired. They do, however, accept it as having some level of historical or theological credibility. There are also questions as to whether the book was actually fully authored by Enoch.

The book of Enoch does identify the seven archangels as Michael, Gabriel, Raphael, Uriel, Raguel, Phanuel, and Sariel. The name Uriel also appears in 2 Esdras, another Jewish text that is not included in the official canon of scripture but is accepted by some Orthodox churches.

The fact that some branches of Christianity accept these extrabiblical names can often lead to confusion. The Catholic Church has been very clear about this issue. It states in the *Directory on Popular Piety*, "The practice of assigning names to the holy angels should be discouraged, except in the cases of Gabriel, Raphael, and Michael, whose names are contained in Holy Scripture." Again, we point out that Raphael is not actually named in the Bible; he is mentioned in the book of Tobit, and only strict Catholics accept two names of God's angels. Any other name is disqualified because it is not part of divine revelation and thus is noncanonical.

Some people develop formal prayers to these specific angels and seek to have them enter directly into their lives. Various occult traditions list the names of many other angels. Accepting these names is an extremely irreverent choice for a Christian since other spiritual beings (much less friendly than the archangels) may answer the call instead. Remember in all this that demons are fallen angels here on earth that seek an invitation or even the smallest opportunity to influence humans and guide them in a direction away from God.

---

[14] Wayne Nalls, "People Who Walked with God," The Journey, July 18, 2018.

There is much in the spirit world that people simply do not know, and that is OK. The church teaches Christians to believe and obey in what we do know, and in that way we can remain faithful to God and the messages given in the Bible.[15]

It is a slippery slope for a Christian (or anybody) to glorify an angel. To put one's faith in a guardian angel and/or praying to a guardian angel is to make that angel an idol in one's life. God is quite clear in His commandments that we must not worship idols. In fact, it is His first and most important commandment: "Thou shalt have no other gods before Me."

Angels are servants of God who obey Him. God can send angels to help and protect us, but these angels are only obeying God's commands. In fact, the Bible tells us that the unfallen angels do not accept worship, so to worship them is a futile exercise. Fallen angels, however, do accept worship—and that is what they actively seek. Therein lies the danger, and we can see it happening all over the world today. We live in very dangerous times in this regard.

Arguments to dispute the fact that there is no such thing as a guardian angel more often than not will also include Jesus's comment in Matthew 18:10, which states, "See that you do not despise one of these little ones. For I tell you that their angels in heaven always see the face of My Father in heaven."

What the verse means is that angels serve humankind by protecting them at God's command. The angels are always looking into the face of the Father in heaven. Angels are not on or visiting earth protecting an individual whenever they so choose. The importance rests on the angel's submission and loyalty to the Father and not so much on his actual duty of providing protection.[16] The point is that the vast majority of humans on earth believe in angels and believe that angels are messengers serving God—and this is a message that we seek in *Angel Communication Code*, a message from the highest form of messengers known to humanity.

How does all of this compare to the percentage of people who believe in extraterrestrials? The PEW Research Center published a report in June of 2021 that found 65 percent of Americans believe extraterrestrials exist and that they are not a threat.[17] The US Office of the Director of Intelligence published a report dated June 25, 2021, that cites remarkably comparable results.[18] The private sector and the military sector have produced independent results on this matter, and now Christian leadership has recently made public that they also believe in intelligent life in the universe that is not of this earth.

Father Emmanuel Carreira, the Vatican's chief astronomer, has stated publicly that there is no conflict between believing in God and angels and the possibility of extraterrestrial brothers that are perhaps more evolved. Father Carreira operates the telescope at the Vatican Observatory in Castelgandolfo, south of Rome.

Reverend Jose Gabriel Funes is a Jesuit priest who manages the Vatican Observatory south of Rome and in Arizona. These observatories are actively seeking other life in the universe and even the origin of the

[15] Philip Kosloski, "The Spiritual Dangers of 'Angel Worship,'" Aletia, September 12, 2019.

[16] "Is It Biblical to Believe That Everyone Has Their Own Guardian Angel?," Christianity.com, November 9, 2021.

[17] Courtney Kennedy and Arnold Lau, "Most Americans Believe in Intelligent Life beyond Earth; Few See UFOs as a Major National Security Threat," Pew Research Center, June 30, 2021.

[18] Office of the Director of Intelligence, "Preliminary Assessment: Unidentified Aerial Phenomena," June 25, 2021.

universe itself. Funes was also a scientific adviser to Pope Benedict. He told Vatican newspaper *L'Osservatore Romano* in an interview, "In my opinion this possibility [of life on other planets] exists. How can we exclude that life has developed elsewhere?" He explained that the substantial number of galaxies with their own planets made this possible. When asked if he was referring to beings similar to humans or even more evolved than humans, he said: "Certainly, in a universe this big you can't exclude this hypothesis. Just as there is a multiplicity of creatures on earth, there can be other beings, even intelligent beings, created by God. This is not in contrast with our faith, because we can't put limits on God's creative freedom. Why can't we speak of a brother extraterrestrial? It would still be part of creation." The church now accepts that the creation of humans cannot exclude the creation of other beings on other planets. They themselves are actively and openly looking for ETs every day.

During the interview headlined "The Extraterrestrial Is My Brother," Funes said he saw no conflict between belief in such beings and faith in God. He held out the possibility that the human race might actually be the "lost sheep" of the universe. "There could be other beings that remained in full friendship with their Creator," he said.

Funes went on to discuss the topic of the big bang theory. Christians have sometimes been at odds with scientists over whether the Bible should be read literally, and issues such as creationism versus evolution have been fiercely debated for decades. Funes said dialogue between faith and science could be improved if scientists learned more about the Bible and the church kept more up to date with scientific progress. He said he believed as an astronomer that the explanation for the start of the universe was "the big bang," the theory that this all sprang into existence from a dense matter explosion billions of years ago. He did clarify this position and said this was not in conflict with faith in God as a Creator. "God is the Creator. There is a sense to creation. We are not children of an accident. As an astronomer, I continue to believe that God is the Creator of the universe and that we are not the product of something casual but children of a good Father who has a project of love in mind for us."[19] This represents a full 180-degree turn in Christian policy given that there was a time in the history of Christianity not so long ago when such statements would be considered heresy and one could be imprisoned, tortured, and burned alive at the stake for making them.

For example, it was only approximately four hundred years ago, on February 16, 1600, when the Roman Catholic Church executed Giordano Bruno, an Italian philosopher and scientist, for the crime of heresy. He was publicly burnt alive at the stake for this crime. Church authorities were fearful of the ideas of this man who was known throughout Europe as a bold and brilliant thinker. Throughout his life, Bruno supported the Copernican system of astronomy, which places the sun, not the earth, at the center of our solar system. He further believed in a universe full of multiple worlds. This sort of thinking was way ahead of its time because there was no defensible proof. He opposed the authority of the church and refused to recant his philosophical beliefs throughout his eight years of imprisonment by the Venetian and Roman Inquisitions. His life is an inspiration to those who are on a quest for knowledge and truth and brought in the amazing period in our history known as the Renaissance, from when so much in classical art, free thought, and new science derives. He died a martyr for the pursuit of scientific truth.[20]

In 1992, after twelve years of deliberations, the Roman Catholic Church reluctantly and finally admitted that Galileo Galilei had been right in supporting the theories of Copernicus. The Holy Inquisition had forced an

---

19 Philip Pullella, "Vatican Scientist Says Belief in God and Aliens Is OK," Reuters, May 13, 2008.

20 Wikipedia, s.v. "Giordano Bruno."

elderly Galileo to publicly recant his ideas under threat of torture in 1633. However, Bruno made no such admission, and therefore he paid a most gruesome price for staying true to his beliefs.

In 2000, the church considered a new batch of apologies. A theological commission headed by Cardinal Joseph Ratzinger was the head of the Congregation for the Doctrine of the Faith, the modern successor of the Inquisition. Ratzinger completed an inquiry entitled "The Church and the Faults of the Past: Memory in the Service of Reconciliation." It proposes making an apology for church errors of the past. Bruno's execution was one of the church's errors that was under consideration, but in Bruno's case, the church made no concessions. His writings remain on the Vatican's list of forbidden texts. The Index Librorum Prohibitorum, also known as the Index of Forbidden Texts, is a very real list of material banned by the Catholic Church. The Index was originally published in the 1500s with the last edition published in 1948. The books and articles listed, of which there are thousands, were banned in order to protect followers from immoral or blasphemous material. In 1966, the church officially ceased the support and use of the Index of Forbidden Texts.[21]

A number of strict Catholic leaders opposed these error investigations from the outset, saying that excessive penitence and self-questioning could undermine faith in the church and its institutions. The current attitude of the Roman Catholic Church toward Bruno is defined by a two-page entry in the latest edition of the *Catholic Encyclopedia*. It describes Bruno's intolerance and berates him, declaring that "his attitude of mind towards religious truth was that of a rationalist." The article describes Bruno's theological mistakes and his lengthy detention at the hands of the Inquisition in detail, but it fails to mention the best-known fact: that the church authorities seized him and publicly burnt him alive on a stake.

Bruno has long been revered in the scientific community as a martyr to scientific truth. In 1889 a monument to Bruno (shown below) was erected in Rome at the location of his execution.

---

21 Loras College Library, "Banned Literature."

Scientists and poets had such strong feelings for Bruno that they paid tribute to him. A book was written detailing his life's work. In a dedication for a meeting held at the Contemporary Club in Philadelphia in 1890, American poet Walt Whitman wrote: "As America's mental courage (the thought comes to me today) is so indebted, above all current lands and peoples, to the noble army of old-world martyrs past, how incumbent on us that we clear those martyrs' lives and names, and hold them up for reverent admiration as well as beacons. And typical of this, and standing for it and all perhaps, Giordano Bruno may well be put, today and to come, in our New World's thankfulest heart and memory."

An examination of Bruno's philosophical legacy reveals a complex figure who was influenced by the various intellectual trends of the time in a period when modern science was just beginning to emerge. His aggressive written assaults on the church earned the admiration of the most advanced thinkers of the period. He also earned the loathing of the church, whose authority was being shaken to the core by other scholarly attacks from the likes of Galileo and other prominent thinkers of the time.[22]

That is one story about Bruno, but there is another, much different story as well. You can look him up on the National Catholic Register ("The Truth about Giordano Bruno" [ncregister.com/blog/the-truth-about-giordano-bruno]) and get the church's version of his history. It paints a very different, much less flattering picture. Both versions cannot be true. This is a perfect example of how history is recorded, as previously discussed in *Angel Communication Code*. The story that is written tells us as much about the author as it does the subject. There are numerous examples. So, what does that tell us about the books of the Bible and the messages from angels within it? It tells us that there is likely much more there than there appears on the surface—if you are looking for it with an open mind.

The evolving church's position in today's world makes the subject of angels and extraterrestrial communication safe and more applicable to *Angel Communication Code*. The Christian church is now a safe harbor on earth for the simultaneous belief in both angels and extraterrestrials. This is not to suggest that the church includes angels in the same category as extraterrestrials, not by any stretch of the imagination. There are, however, those who do in fact believe that angels are extraterrestrials by pure definition. The point being that most everybody believes in angels. They believe and understand that the purpose of angels is to be messengers. We seek their message in *Angel Communication Code* as it relates to a link to establishing extraterrestrial communication. Precisely and completely what angels are remains an openly debated question.

There is a faction people out there who claim angels are extraterrestrials and extraterrestrials are angels. All angels are extraterrestrials, but not all extraterrestrials are angels, right? This premise is the raw definition of extraterrestrials as intelligent beings that do not dwell on Earth. There is much more to it than that. The words *angel* and *extraterrestrial* cannot and should not be used interchangeably. Angels dwell in heaven, and heaven is not a habitable planet. Angels are purely spirit and do not have biological form. Angels are not another species of ET and are not akin to any of the other ET species that might exist out there in the far corners of the universe. Is that really the case, though? Does an extraterrestrial have to be made of flesh and blood? This calls into question the possibility of life in other dimensions versus on other planets in the universe. We can send a rocket to another planet that may support extraterrestrial life, but we cannot send a rocket to another dimension or to heaven, where God, the angels, and a saved human spirit can dwell after human biological life has expired.

22 Frank Gaglioti, "A Man of Insight and Courage," International Committee of the Fourth International, February 16, 2000.

There are many written publications and other venues out there produced by reputable people still trying to sell us on the notion that ETs are out there and that they have visited Earth and influenced the development of humanity. This subject has been examined and has been written about over and over again. There are television documentaries, lecture circuits, website groups, and books that try to argue those specific types of questions. Taking a position on that subject here in *Angel Communication Code* is not germane to the subject matter of this book. This book does not question the existence of intelligent life in the universe not of this earth. This book is about communication with ETs, which we know exist, and clues to a communication code that may have been provided by ETs or angels or both. Establishing communication with intelligent life that does not dwell on Earth is our focus. *Angel Communication Code* admits the fact that ETs exist without hesitation or reservation.

At this point, the message has been received loud and clear. We get it. ET visitations happen, UFOs can come and go at will, and abductions have occurred, so let us stop beating that subject to death and move on to the next step. Researcher needs to dig deeper to address what should be the next step, which is establishing communications with ETs. There is a difference between contact and communication. Let us move forward and figure out how to achieve what is not apparently happening, which is two-way communication with our ET brothers and sisters in modern times.

Why is the key to establishing this communication buried in clues and not just given to us? Why must there be a message to find at all? Perhaps it is because we have to prove ourselves worthy of establishing ET communication by finding the communication code before we can earn the big reward of being recognized as worthy members of the universal community. Demonstrating worthiness to receive something so grand is not an uncommon concept. Humans do this with animals all the time, measuring the intelligence of a subject animal in behavioral research experiments.

Why make the mouse find its way through a maze to get to the cheese? We do it just to see if the mouse can. That is a very basic example of a reason for scientific research. Using animals for scientific research has been a controversial practice for a long time. Much human benefit has resulted from this type of research at the expense of animal health or life. Is it entirely out of the realm of possibility that humans are the animal of choice that ETs use for behavioral studies? There are dozens of alien abduction reports available in which the abductees claim they were subject to some form of terrifying medical examination or probe. There are also numerous examples of humans being taken alive to visit heaven or the throne of God for the purpose of educating these individuals. Humans have taught many animal species to perform tasks that benefit humans and not so much the animal. Take service dogs, for example. Could that be the ETs' plan for us? If so, we must be very slow learners. The ET influence on humanity has been going on for thousands of years.

Medical research is a valid reason to abduct any human (or other animal). Humans do the exact same thing to all sorts of earthly creatures. Behavioral research is the same thing. We make the subject animal, as an individual or in groups, follow the clues to the prize so we can learn things about the animal and perhaps ourselves. It becomes quite unsettling if we start to think about the dynamic between humans and ETs from the perspective of a lab rat.

Now consider human entertainment such as gambling on the performance of animals for no other purpose than our own amusement. Whether it's animals racing to a finish line or fighting to the death, humans have been gambling on the outcome of animal performances in this earthly arena for centuries. Perhaps humans are just pawns in the very same game of chance in the arena of the universe, purely for the amusement of

ETs. Why not? We are tiny in the universe and are without the ability to leave our earthly cage under our own power, with the exception of a few astronauts for a short time and distance. In essence, humans are in a cage we cannot escape, and ETs do not even have to feed us or shelter us to keep us alive and available to them. It is really a matter of circumstance and perspective.

Imagine being a squid living deep in one of the oceans on earth. The deep ocean is the only world you know. Then one day you are snagged in a net; are hauled to the surface, where you experience sunlight for the first time; and are chucked out onto the deck of a fishing boat and can't breathe but can see all sorts of alien things in that new world. Next thing you know you are the calamari appetizer on the menu of that new world. Is that an ET experience for the squid even though the squid never left Earth? It is a matter of circumstance and perspective.

The creatures we use for research or for our own recreational purposes are in a subcage until we choose to use them. We keep these creatures in their own private cages of our design until we put them into an arena with a competitor for survival so we can gamble on their performance in a gruesome fight to submission or death. It is a horrible but very real thing in our world.

Is it so far-fetched to consider that the possibility exists that ETs are using humans for medical research, behavioral research, appetizers, or just their own amusement in games of chance? Perhaps the ETs will recognize and openly engage with humans, and not use humans, when we find the code to two-way communication and demonstrate the ability to use it. Again, we can draw a parallel to this between humans and earthly creatures. Humans treat dogs and dolphins differently than they treat insects because we recognize the dog's and dolphin's ability to communicate with us. We consider them intelligent. Is identifying the ET communication code analogous to finding the cheese at the end of the extraterrestrial maze in some way? The game might be for us to find the code and establish two-way communication in order to survive. The grand prize is entry into the universal community of intelligent life or even eternal life in heaven. Get your head around these concepts and try hard to expand your mind to embrace that level of possibility—because it could be quite real. This reward concept is not new. There are many examples. One example is the twelve labors of Hercules from Greek mythology.

The short story goes something like this—notice the relevant symbolism:

Hercules was the son of Zeus, king of the gods, and Alcmene. Zeus took on the form of Alcmene's husband, Amphitryon, and visited Alcmene one night in her bed, and so Hercules was born a demigod with incredible strength and stamina. Hercules performed amazing feats of heroism during his life, including wrestling death and traveling twice to the underworld. His life was far from easy from the moment of his birth. This was because Hera, the wife of Zeus, knew that Hercules was her husband's illegitimate son. She wanted him destroyed. Hera coerced Hercules into killing his wife, Megara, along with their children. When Hercules realized what he had done, he deeply regretted it and went to the Oracle of Delphi to ask for penance.[23] Hercules was directed by the Oracle of Delphi to serve Eurystheus, king of Tiryns, for twelve years. If Hercules was able to complete all the required tasks (proving himself worthy), he would be given immortality—the greatest of all gifts.

The belief in eternal life has a familiar ring today in that most people of faith seek eternal life in their respective equivalent to heaven. Is it myth, or is there something to the achievement of eternal after life via

---

[23] *World History Encyclopedia*, s.v. "The Life of Hercules in Myth and Legend," Joshua J. Mark, July 23, 2014.

a demonstration of worthiness? The bottom line is that unless we gain some hard and factual knowledge, we will never know until we die. Faith is a much more complicated thing when we bring ETs into the same room with religious faith.

The entire concept of Christianity and other faiths is about living one's life in a way that will earn one entry into heaven or some other version of an afterlife for the soul and spirit after death. There are numerous biblical examples of someone proving their worthiness to receive something grand. David slew Goliath and proved he was worthy to one day be king. Abraham was willing to sacrifice his son, as directly ordered by God, to prove his faith and worthiness to be received into the kingdom of God. He was stopped at the last instant by a message from God delivered by an angel. There are many other examples. We can compare how humans use animals to how ETs could be using humans, but there is a difference. Humans are guided by messages delivered by God through angels, and that makes us unique. Lab rats, not so much.

Angels and ET beings are not one and the same or even similar, but perhaps they are working either independently or together to give us the same information, the information that will allow us humans to prove ourselves worthy of that grand prize of acceptance into the universal community beyond Earth. Many books of the Bible reference angels as previously discussed in *Angel Communication Code*. The questions are what (or who) exactly are these heavenly creatures, what role did they play in ancient history, what role do they play in the modern world, and did they leave us clues to a code for establishing two-way ET communication?

Mary, the human mother of Jesus Christ, who is the only begotten Son of God, is often referred to as "queen of the angels"; however, she did not die and was not given angel status by God in heaven. That has never happened to any human, including Mary, as far as we know. The notion of humans becoming angels is exactly that—a human notion. It is not biblical. The notion of Mary being the queen of angels is something that was conjured up by Saint John of Damascus and his interpretation of Colossians 1:16: "For in Him, were all things created in heaven and on earth, visible and invisible, whether thrones or dominations, or principalities or powers; all things were created by Him and in Him."

Because God the Father, Jesus His Son, and the Holy Spirit are one and the same, according to Saint John of Damascus this means that Mary is, by default, the queen of the angels. That is a long stretch for sure, but this is believed to be where the notion of Mary as queen of angels was hatched.

Mary is said to be able to perform as a messenger who can deliver a human message to God, but not the other way around like an angel. She cannot directly deliver on any prayer request but can only deliver a message to God on a person's behalf. Mary does not deliver messages from God to people the way angels deliver messages.[24] Praying to any deity other than God is a sin, but like all sins, it is forgivable by God if the sin is recognized, understood, confessed, and repented.

Consider the Catholic Hail Mary prayer, which has a very long and complex history. This prayer is not about praying to Mary per se. It is about asking Mary to help a person pray to God for forgiveness of our sins: "pray for us sinners now and at the hour of our death." The prayer is asking Mary to help us reach God while we are still alive and making final preparations to be judged for worthiness of eternal life in heaven. The Hail Mary prayer is not part of Christian Eucharistic worship and services as is the Lord's Prayer, which is often

---

[24] "Mary Queen of Angels," Catholic Harbor of Faith and Morals, accessed November 24, 2023.

mistakenly referred to as the Our Father. The Lord's Prayer is a prayer directly aimed at God without help from Mary or any divine entity asking Him directly to "forgive our trespasses."

The Hail Mary prayer is part of the rite of Christian confession, after which the priest directs the confessor to recite a variety of prayers numerous times in series. Typically, these would include both the Lord's Prayer and the Hail Mary prayer.

The Hail Mary prayer is also a large part of the rosary prayer sequence. The idea of the rosary was provided to Saint Dominic, a Dominican priest, when—as it is said—the Virgin Mary appeared to him as an apparition in 1214. Saint Dominic was a priest and was the theologian Alanus de Rupe, who encouraged the practice of the rosary by defining the fifteen rosary promises and founding several rosary confraternities, which are Christian voluntary associations of ordinary people created for promoting special works approved by church leaders.

The rosary is recited in honor of the Virgin Mary. It is made up of a set number of specific prayers recited in a specific sequence. These prayers include the Apostles' Creed, the Lord's Prayer, the Hail Mary, and the Glory Be. The purpose of the rosary is to maintain and memorialize specific and fundamental biblical principles as those principles are portrayed in the history of our hopes of salvation and eternal life in heaven. There are twenty mysteries to reflect upon while praying through the rosary. Those mysteries are separated into the following:

- Five Joyful Mysteries (said on Monday and Saturday),
- Five Luminous Mysteries (said on Thursday),
- Five Sorrowful Mysteries (said on Tuesday and Friday), and
- Five Glorious Mysteries (said on Wednesday and Sunday).

Properly completing the rosary prayer is not a quick or casual thing. It is a commitment.[25] One must pray his or her way around the string of beads one by one in a specific order. There are fifty-nine beads on a rosary.

All this attention, honor, and reverence to Mary the mother of Jesus the Christ is very important to the Christian faith, and Mary's stature is beyond reproach; however, there are no biblical references to Mary becoming an angel in heaven after her human, earthly death, nor is she part of the accepted angel hierarchy defined by humans.

Much is revealed in scripture about God's directive to angels, their appearances, and their capabilities. As noted previously, one important yet basic question about angels that remains unanswered is about precisely when and how angels were created or otherwise came into existence. Scripture speaks about the creation of humankind but not about the creation of angels. It is unclear if God created the angels as part of creating the universe or if the angels were at His side already when He created the universe. If you're a big bang believer, the question is the same. Did the big bang happen first to create the environment necessary for the creation of God and His subsequent creation of angels, or were angels already in existence when the big bang happened? Was it God who created the universe via the big bang? It really does not matter, because in either scenario, God delivers His messages to humans through angels. We need to interpret and decipher those messages at more than face value to get to the ET communication part of the program.

---

25 "How to Pray the Rosary," Rosary Center and Confraternity.

Given the importance of angels as messengers and servants of God throughout the Bible, the lack of consistent direct information about their creation and hierarchy is important to note. One would expect that the Bible would be definitive and clear on this matter, but it yields only vague, mixed, and contradictory information about angels. Was this done intentionally for some reason known only to the authors of the books of the Bible? It is these ambiguities that open the door for a closer look at angels with a focus on ET communication and the deposition of clues to an ET communication code.

Because the Bible avoids a precise explanation or details about the creation of angels, Christian clergy and theologians had to develop and agree on their own definition(s), which are based on their interpretations on what little information is offered about the creation and nature of angels throughout the Old and New Testaments and in other Christian texts as well. Before we can discuss any links between an ET communication code and angels, we must first understand what angels are all about in this context.

The most basic definition of angels developed by theologians and church clergy during the early years of Christianity is, as we have already touched upon, that they are spiritual beings created by God to serve God. Other words are also used for these spiritual beings.[26] If you look it up in a dictionary, you will find that a synonym for the word *angel* is "supernatural being," and a synonym for *supernatural being* is "alien," and a synonym for *alien* is "extraterrestrial." The words *angel* and *extraterrestrial* are clearly linked.

The primary word used for spiritual beings in the Bible is *angel*. As noted previously, the Hebrew word for angel is *malach*, and the Greek word is *angelos*. Both words mean "messenger," so let us dissect that word a bit. A messenger is defined as one who executes the purpose and will of the one whom he or she serves. We have learned thus far that good angels are obedient to God and carry out His will, while the fallen angels disobeyed God and fell (meaning that they were pushed out by God) from their holy position. Fallen angels exist in active and perpetual opposition to the work and plan of God.[27] There is no return to grace for a fallen angel, and angels never die—and no new angels are ever created as far as we know. To the best of our knowledge, angels do not reproduce in the way humans reproduce to make baby angels that look like plump baby cherubs as they are often depicted in some works of art.

Satan himself is the fallen angel who was originally the good angel named Lucifer. Christians believe Lucifer became associated with the name of Satan just before he fell from heaven. The Catholic Church has always taught that Satan was at first a good angel made by God. The devil and the other demons were initially created to be good by God, but they became evil by their own doing.[28] Various traditions claim that Lucifer was the brightest among the angels before he rebelled against God.[29]

The name Lucifer is complex and has morphed over the centuries. *Lucifer* is a Latin word that means "light-bearer" and was originally the name for the planet Venus, known as the "morning star" or "brightest star." In the Old Testament, the planet Venus is referred to using this name. Jesus also refers to Himself as the morning star in Revelation (22:16): "I am the root and the descendant of David, the bright morning star." So here we have two stars using the same star reference to identify themselves. One star is still shining, and the other star has fallen. This dynamic is the root of the battle between two sons of God: Jesus, His only

---

26 "Angelology: The Doctrine of Angels," Bible.org.

27 Ibid.

28 Catechism of the Catholic Church, chapter 391.

29 Philip Kosloski, "Who Is Lucifer and What Does His Name Mean?," Aleteia, September 12, 2019.

begotten Son, and Satan, His first created son. Satan was originally the brightest star, but his light has been extinguished. Satan, however, was not a begotten son. Jesus, the only begotten Son of God, is now the brightest star still shining. Although He was executed and sacrificed in exchange for the forgiveness of all human sin, He rose from the dead and assumed His place in heaven at the right hand of God. He Himself then became God via the concept of the Holy Trinity. His light was extinguished for a short time but would not be permanently extinguished.[30] There are those who contend that Jesus is the King angel, the brightest angel among the stars. Others contend He is separate from the angels entirely.

In Exodus 4 in the Old Testament, the human Israelites are referred to as the sons of God; however, the first nonhuman beings to be called sons of God (in the book of Job) are the angels. When the early Christians were traveling and preaching the good news of the birth of Jesus to Jewish and Gentile audiences, they would say that Jesus is the Son of God. They did not mean to imply that He is one of the angels. In the book of Hebrews it says: "He is as much superior to angels as the name he has obtained is more excellent than theirs" (1:4). It goes on to say: "For to what angel did God ever say, 'Thou art my son, today I have begotten thee?'" (1:5). The very definition of *begotten* is "to be brought into existence by or as if by a parent, or to bring a child into existence by the process of reproduction."

The angels are all created sons of God as cited. Jesus, however, is the only begotten Son of God, which confirms that He is different and of a higher status than the angels. The only begotten Son of God, Jesus, was not created in the same way as angels. He is eternally begotten of the Father. He was born the flesh and blood Son, not an angel. He always has been that only Son. He always will be that only Son. He is the Son of God the Father, through whom the entire universe and all its creatures (including the angels) were made.[31] That is how the Christian faith is built. We can only wonder about the spiritual faith of ETs. What do they believe in, if anything, spiritual? How do they suppose that their world and they themselves came into existence? Until we can communicate with them and have these debates with them, we will never know.

Long before there was Jesus, there were angels. God did not decide to sire a child with a human woman until He observed the behavioral trends of His first human creations. It is not unreasonable, but perhaps blasphemous, to suggest as a faithful Christian that Christianity's all-powerful and all-knowing God of our creation is or was, in fact, not completely all-knowing and perfect as Christians are taught and are required to believe. A scientist understands that nothing we know about the universe or faith of any denomination is exempt from scrutiny.

The raw facts are that (1) Christian faith tells us that God created humans and angels and (2) His creations fell short of His expectations, requirements, and intent, so He destroyed the unsatisfactory ones. Pure logic tells us that even if God is perfect, He was capable of a mistake. If all He ever did and does were perfect, then the destruction of His imperfect creations would have been unnecessary. His creation would have been perfect from the start. His angels and humans would have been obedient right out of the gate.

God's gift of free will is the wildcard in all this scrutiny of His perfection. God gave his angel and human creations free will. All are capable of disobedience and disbelief. Apparently, God did not control everything, and if God is perfect, then we must assume this was His intent. God's intent and expectations for His human creation was that they would choose obedience and faith. The other option is death. Seems like an easy

[30] Candice Lucey, "Why Are Both Jesus and Satan Referred to as the Morning Star?," Christianity.com, October 7, 2020.

[31] Brant Pitre, "Is Jesus an Angel? Hebrews 1, the Incarnation, and Jehovah's Witnesses," Catholic Productions, December 27, 2019.

choice. Free will and the influence of the enemy (Satan) interfere with the expectations of faith. Just look at the disturbing things going on in the world today. It is terrifying. It is all playing out as prophesied in scripture.

God is not human, and perhaps He determined that He needed a half-human Son so He could experience a much closer, direct, and intimate human experience. Through the experiences of His half-human Son, Jesus, God would learn much about His human creation. Only through the creation of a flesh and blood half-human Son could God truly understand His human creation. He initially made a questionable choice in giving people free will to choose between disobedience and obedience. Jesus provided a solution to that problem and a means of forgiveness for humans falling short of God's requirements as cited in the Ten Commandments. Only through the experiences of Jesus could God truly understand His creation, and that is a legitimate proposal for why God created Jesus. Implying that God is perhaps not 100 percent perfect is a dangerous part of what we are focused on in *Angel Communication Code*, going back to basics and thinking differently.

In the hierarchy of holy beings in the centuries before there was Jesus, God was number one, and then came the angels, and after that were the humans. There is a hierarchy of angels, and only the top-tier group of them has direct access to God. Lower-tier angels must work through the top-tier angels to get to God. The lowest level of angels is called "angel" and is just above humans. Jesus would be the exception. Jesus was the half-human Son of God and had direct access to His Father. Jesus did not have power over angels while He was in the flesh. In His glorified status after His resurrection, He was lifted to a position of authority well above the angels. The apostle Paul states four distinct times in Ephesians 1:20–22 that after Jesus's resurrection, God lifted Him to a position of authority over all creation, including the angels. Now the angels are subject to His command as the number one holy entity. Jesus, God, and the Holy Spirit are now one and the same. This trilogy or triad is one of the most difficult concepts to grasp in Christianity. This means that before His resurrection, Jesus was God's messenger with direct access to God in His (Jesus's) human form. Any sort of instructions as to a way for future humans to leave earth and get to heaven were communicated through Jesus when He was in human form. Angels, of course, were also delivering messages to humans while Jesus was alive. Jesus did not replace angels in that capacity. After the resurrection, Jesus became God's messenger in spirit form like the angels, effectively making Him the most important of all God's messengers. There is still to this day debate among some theologians and church clergy as to whether the resurrected Jesus is or is not an angel by pure definition.

What is the relationship between Jesus and the angels with respect to the subject of *Angel Communication Code*? Whether the resurrected Jesus is or is not considered an angel is irrelevant. What matters is the fact that Jesus is the King of angels beyond any contestation, and because of His status, the messages delivered by angels come from Him as God, the Holy Spirit, and King of angels, making any clues to a communication code derived from His words and deeds relevant to *Angel Communication Code*.

The Bible clearly indicates that not all angels are the same. These spiritual beings are different in appearance and were created by God to perform a variety of missions, depending on to which angel choir a particular angel belongs. Bible references that include angels often refer to specific types of choirs. Archangels, cherubim, seraphim, and others are all choirs of angels referenced in the Bible. The Bible, however, never completely categorizes all the different choirs of angels into any formal ranking or importance or hierarchy. The Bible does make it obvious that some angels are of higher status than others and that each choir has a specific roster of tasks that they have been selected to perform for God. Sometimes the lines are blurry and contradictory in scripture, but the lines are implied.

It was during the Middle Ages when the church decided there was a need for the formal delineation of the angel hierarchy. Perhaps the reason was simply to recognize that not all angels were created equal, or perhaps there was much more to it than that. A variety of angel order structures proposed by different people were being considered by the church. One blueprint for the structure of angels was proposed by a person identified as Pseudo-Dionysius, a mystic. Others were suggesting completely different angel hierarchies.

Dionysus (sometimes spelled Dionysos) was the Greco-Roman nature god of fruitfulness and vegetation, more specifically considered as the god of wine and ecstasy. His name in Linear B script (developed in the thirteenth century BCE) shows that he was already worshipped during the Mycenaean period (1750 BC to 1050 BC). The origin of his cult is not specifically known; however, in all the legends of his cult, he is described as having foreign origins.[32]

The term *dionysius* is a bit strange and was used quite a lot in ancient times. Etymologically speaking, the term is a nominalized adjective formed with an *-ios* suffix from the root *dionys*, which is derived from name of the Greek god Dionysus. The exact reasons for assignment of such names are not specifically known. *Dionysios* itself refers only to males. The feminine version of the name is Dionysia in both Greek and Latin. *Dionysius* is sometimes used as a title in a religious contexts as well. *Dionysius* was the episcopal title of the founders of the Malankara Church (founded by the apostle Thomas in India) from 1765 until the synthesis of that title with Catholicos of the East in 1934.[33]

This mystic identified only as Pseudo-Dionysius developed a structure of three angel hierarchies, spheres, or triads, with each hierarchy containing three orders or choirs of angels.[34]

Pseudo-Dionysius the mystic was a Christian Neoplatonist. *Neoplatonism* is a more contemporary term used to describe the period of Platonic philosophy beginning with the work of Plotinus and ending with the closing of the Platonic Academy by the emperor Justinian in 529 CE. This type of Platonism is described as mystical or religious in nature, developed outside the mainstream of Academic Platonism. The origins of Neoplatonism can be traced back to the era of Hellenistic syncretism, which ignited movements and schools of thought such as Gnosticism and the Hermetic tradition. A major factor in this amalgamation of religious philosophies, which had an immense influence on the development of Platonic thought, was the inclusion of Jewish scriptures into Greek intellectual circles by way of a translation called the Septuagint, which is a blending of the creation narrative of Genesis and the cosmology of Plato's *Timaeus*. It was the start of a long tradition of cosmological theorizing that led up to the publication of Plotinus's *Enneads* (a collection of nine writings).

The work of Pseudo-Dionysius is a complex synthesis of Platonic philosophy and Christian theology that went on to become a huge influence on medieval mysticism and Renaissance humanism.[35]

---

32 *Encyclopedia Britannica*, s.v. "Dionysus / Greek Mythology."

33 Wikipedia, s.v. "Dionysius."

34 Ralph M. McInery, ed., *Selected Writings of Thomas Aquinas* (New York: Penguin, 1998), p. 841.

35 Internet Encyclopedia of Philosophy, s.v. "St. Elias School of Orthodox Theology," Edward Moore.

Pseudo-Dionysius transformed from a pagan Neoplatonism way of life into a distinctively new Christian way. His book *On the Celestial Hierarchy* includes the following chapters:[36]

- Chapter 1. That every divine illumination, whilst going forth lovingly to the objects of its forethought under various forms, remains simplex. Nor is this all. It also unifies the things illuminated.
- Chapter 2. That divine and heavenly things are appropriately revealed, even through dissimilar symbols.
- Chapter 3. What is hierarchy and what the use of hierarchy?
- Chapter 4. What is meant by the appellation "angels"?
- Chapter 5. For what reason all the heavenly beings are called, in common, angels.
- Chapter 6. Which is the first order of the heavenly beings? which the middle? and which the last?
- Chapter 7. Concerning the seraphim and cherubim and thrones, and concerning their first hierarchy.
- Chapter 8. Concerning lordships and powers and authorities, and concerning their middle hierarchy.
- Chapter 9. Concerning the principalities, archangels, and angels, and concerning their last hierarchy.
- Chapter 10. A repetition and summary of the angelic discipline.
- Chapter 11. For what reason all the heavenly beings, in common, are called heavenly powers.
- Chapter 12. Why the hierarchs amongst men are called angels.
- Chapter 13. For what reason the prophet Isaiah is said to have been purified by the seraphim.
- Chapter 14. What the traditional number of the angels signifies.
- Chapter 15. What are the morphic likenesses of the angelic powers? what the fiery? what the andromorphic? what are the eyes? what the nostrils? what the ears? what the mouths? what the touch? what the eyelids? what the eyebrows? what the prime? what the teeth? what the shoulders? what the elbows and the hands? what the heart? what the breasts? what the back? what the feet? what the wings? what the nakedness? what the robe? what the shining raiment? what the sacerdotal? what the girdles? what the rods? what the spears? what the battle-axes? what the measuring lines? what the winds? what the clouds? what the brass? what the electron? what the choirs? what the clapping of hands? what the colors of different stones? what the appearance of the lion? what the appearance of the ox? what the appearance of the eagle? what the horses? what the varieties of colored horses? what the rivers? what the chariots? what the wheels? what the so-called joy of the angels?

His justification for the development of an angel hierarchy is described as follows, taken directly from his book:

> Moreover, every divine procession of radiance from the Father, while constantly bounteously flowing to us, fills us anew as though with a unifying power, by recalling us to things above, and leading us to the unity of the Shepherding Father and to the Divine One. For from Him and into Him are all things, as is written in the holy Word.
>
> Calling then upon Jesus, the Light of the Father, the Real and True, "Which lights every man that comes into the world, by whom we have access to the Father," the Origin of Light, let us raise our thought, according to our power, to the illumination of the most sacred doctrines handed down by the Fathers, and also as far as we may let us contemplate the Hierarchies of the Celestial Intelligences revealed to us by them in symbols for our upliftment: and

36 *Stanford Encyclopedia of Philosophy*, ed. Edward N. Zalta (Stanford, CA: Stanford University, Spring 2019), "Pseudo-Dionysius the Areopagite."

> admitting through the spiritual and unwavering eyes of the mind the original and super-original gift of Light of the Father who is the Source of Divinity, which shows to us images of the all-blessed Hierarchies of the Angels in figurative symbols, let us through them again strive upwards toward Its primal ray. For this Light can never be deprived of Its own intrinsic unity, and although in goodness It becomes manyness and proceeds into manifestation for the uplifting of those creatures governed by Its providence, yet It abides eternally within Itself in changeless sameness, firmly established in Its own unity, and elevates to Itself, according to their capacity, those who turn towards It, uniting them in accordance with Its own unity. For by that first divine ray we can be enlighted only insofar as it is hidden by all-various holy veils for our upliftment, and fittingly tempered to our natures by the Providence of the Father.

Pseudo-Dionysius would come to be compared to and referred to as Dionysus the Areopagite. For an unidentified person and unknown author and mystic to be equated to Dionysus the Areopagite, or a member of the Council of the Areopagus, tells us that whoever or whatever group chose that name had a lot of respect for his intellectual abilities and mystical gifts. The Areopagus in Athens was the site of a council that served as the highest legal institution under the Athenian democracy. This group, as just identified, was called the Council of the Areopagus. The council existed long before democracy, and its powers and composition changed many times over the centuries. Originally, it was the central governing body of Athens, but in a democracy, it became primarily the court with authority over cases of homicide and certain other serious crimes. After an Athenian had served as one of the nine archons, his conduct in office was investigated, and if he survived that investigation, he became a member of the Areopagus. Tenure was for life.[37]

The definitive hierarchy of angels finally accepted by the Christian church was ultimately the one put forth by Pseudo-Dionysus, whose identity was and remains unknown. What we do know is that he was a fifth-century theologian and mystic. Some believe he was actually a Greek monk. The fact that his true identity was never exactly known is an important matter to the subject of *Angel Communication Code*. Much has been written about this man. The fact that he is referred to as a mystic in most references is even more important to the things we are talking about in *Angel Communication Code*.

A mystic by definition is "a person who claims to attain, or believes in the possibility of attaining, insight into mysteries transcending ordinary human knowledge, as by direct communication with the divine or immediate intuition in a state of spiritual ecstasy."[38]

A mystic whose true name and origin was and is unknown put together and defined the angel hierarchy that was accepted by the church and still is to this day. Not a Bible author, not an apostle, and not a saint, pope, bishop, or cardinal, he was an anonymous mystic, a very complex thinker, and a philosopher whom the entire Christian world accepted and believed from top to bottom.

How is it possible that this mystic was anonymous yet responsible for such a critical responsibility in Christianity? That seems strange at face value, but in a way, it makes perfect sense. If this mystic had "insight into mysteries transcending ordinary human knowledge, as by direct communication with the divine or immediate intuition," then he may also have had knowledge of ETs and how to communicate with them.

---

[37] Christopher Blackwell, "The Council of the Areopagus," *Dēmos: Classical Athenian Democracy* (January 26, 2003).

[38] Dictionary.com, s.v. "mystic".

Weaving clues to that code into his structure of angels, God's messengers, was a golden opportunity for this mystic to use angels to lay down some clues for future, more technologically advanced humans to find.

Pseudo-Dionysus created an angelic hierarchy consisting of three triads, each holding three angel choirs.[39] Right out of the gate, at the core of the angel hierarchy he laid out three triads of three angel types or choirs. This means that each triad holds three of the nine choirs, or three-ninths of the total choirs. The ratio 3:9 reduces to 1/3, or in decimal form 0.33333333, with 3 as an infinite decimal. Recall that this is the very same ratio that describes how one-third of all God's angels fell from grace with Lucifer (Satan). This is another direct, obvious, powerful, and easily identified link to the number 3 of the ET communication code. However, if you're not looking for it, you won't make the connection. This ratio analysis becomes even more interesting when we consider all the ratios of 9 up to the number 9, as follows:

> 1:9 = 0.11111111111111111111111111 to infinite decimal places
> 2:9 = 0.22222222222222222222222222 to infinite decimal places
> 3:9 = 0.33333333333333333333333333 to infinite decimal places
> 4:9 = 0.44444444444444444444444444 to infinite decimal places
> 5:9 = 0.55555555555555555555555555 to infinite decimal places
> 6:9 = 0.66666666666666666666666666 to infinite decimal places
> 7:9 = 0.77777777777777777777777777 to infinite decimal places
> 8:9 = 0.88888888888888888888888888 to infinite decimal places
> 9:9 = 1.0 (the end)

Of course this is just arithmetic. However, nine, the number of angel choirs, points us to infinity. It is unique in that when it is used as the denominator in a ratio, with numerators up to itself, the decimal places are infinite and the numerator of the ratio in each case (7/9 = 0.777, e.g.). It points to infinity, or perhaps eternity in each choir. Pseudo-Dionysus chose to identify nine choirs for a reason that may not be obvious. He did not get it out of the Bible, because it does not exist in the Bible.

The Pseudo-Dionysus angel hierarchy looks like this:

1. First hierarchy
   a. seraphim
   b. cherubim
   c. thrones
2. Middle hierarchy
   a. dominions
   b. virtues
   c. powers
3. Last hierarchy
   a. principalities
   b. archangels
   c. angels

39 "The Celestial Hierarchy of Pseudo-Dionysius the Areopagite," Dallas Baptist University, https://www3.dbu.edu/mitchell/celestialhierarchy.htm.

Pseudo-Dionysus's detailed descriptions of the nine choirs are provided in the next chapter. The nine-count of angel choirs can be linked to other religions, faiths, and myths from all over the world that also have nine-count deities, for example:

Nine Muses

According to classical mythology, the Nine Muses are the daughters of Zeus (father of Hercules) and Mnemosyne (Greek goddess of memory). The ancient Greeks and Romans believed that these goddesses granted inspiration to those involved in creative work such as poets, musicians, and artists. Therefore, the Nine Muses were considered to be the personification of the literary arts, music, and the visual arts, and each of them oversaw a particular aspect of the arts. Today, the domain of the Muses has been extended to include all aspects of art, literature, and also science. This is reflected in the word *museum*, derived from the Greek *mouseion*, which originally meant "temple or shrine of the Muses."[40]

Nine Gods of Mayan Civilization

The nine Mayan gods explain natural phenomena, the origin of humanity, the structure of the universe itself, and everything otherwise inexplicable. In addition, through various discoveries, it can be affirmed that this Mayan religion was used by the rulers as an instrument of control and legitimization of their divine origins before the Mayan people. The Mayan gods were part of a complicated and diverse vision on the part of the Mayan people. Many of these gods presented themselves in different forms depending on the place and time in which they were located. Even their decisions depended on their state of mind, which was directly associated with the number of offerings and sacrifices presented to them.[41]

Egyptian Ennead (Nine) Gods

Ennead translates literally into "nine." Ancient Egyptians created several enneads, or groups of nine gods, as their unification under dynasty. This brought numerous local cults into contact with one another. Ancient Egyptian mythology had many different explanations for the same phenomenon. This concept is especially unique because no single story was more accurate than another was, but the truth was a mix of them all.

The Pyramid Texts of Dynasties V and VI mention the Great Ennead, the Lesser Ennead, the Dual Ennead, and the Seven Enneads.[42]

Chinese Emperor Gods

The Nine Emperor Gods of China are the nine sons manifested by Father Emperor Zhou Yu Dou Fu Yuan Jun and Mother of the Big Dipper Dou Mu Yuan Jun, who holds the Registrar of Life and Death. According to Reverend Long Hua, a thirty-fifth-generation Taoist priest from Singapore, honoring the Northern Dipper stars prolongs one's life, eliminates calamities, and absolves sins and past debts of oneself and one's family.

---

[40] DHWTY, "The Nine Muses: Daughters of Zeus and Memory, Goddesses of the Arts," Ancient Origins, April 6, 2020.

[41] "Mayan Gods," Mayan Peninsula, 2022.

[42] Wikipedia, s.v. "Ennead."

The Nine Emperor Gods Festival celebration is an important ceremony to invoke and welcome the Nine Emperor Gods. Since the arrival of the Nine Emperor Gods is believed to be their descending through the waterways, processions are held from temples to the seashore or river to symbolize this belief. During this period of time, the constant tinkling of a prayer bell and chants from the temple priests or mediums are heard. Most devotees stay at the temple, eat vegetarian meals, maintain celibacy, and continuous chant in prayer. The ninth day of the festival is its climax, which includes an important procession that draws thousands of devotees to send the Nine Emperor Gods back via the waterways.[43]

Hindu Navadurga

For Hindus, the mother goddess Durga is able to appear in nine different forms, each of which is endowed with unique powers and traits. Together, these nine manifestations are called *Navadurga* (translated as "nine Durgas"). Durga is a form of Shakti. The evolution of Shri Maha Saraswati, Shri Maha Laxmi, and Shri Mahakali (the three main forms of Shakti) took place from Shri Brahma, Shri Vishnu, and Shri Mahesh, respectively. Each of these three deities gave rise to three more forms; hence, in all, these nine forms together are known as Navadurga. Notice that this is very similar to the three triads of three types of Christian angels.

Devout Hindus celebrate Durga and her nine denominations during a nine-night festival called Navaratri, which is held in late September or early October, depending on where it falls on the Hindu lunisolar calendar. Each night of Navaratri honors one of the mother goddess's manifestations. Hindus believe that worshipping Durga and celebrating with prayer, song, and rituals with sufficient religious intensity will lift the divine spirit and fill them with renewed happiness.[44]

There are other examples, but the point here is that there are several common threads that weave through religious faiths across the globe going back to ancient times with respect to deity rosters (among several other commonalities). These common threads are all clues that link to communication with beings not of this earth. In *Angel Communication Code* we are focusing on Christian angels, trying to identify their contribution to establishing two-way communication with ETs; however, these links seem to have a global overlap.

What is an angel? An angel is a divine messenger not of this earth.

What do we seek? We seek clues to a message telling us how to communicate with beings not of this earth.

If we seek a place to find clues leading to a way to communicate with ETs, then the divine messengers we call angels would be a logical place to look—with a high probability of success.

---

43 Wikipedia, s.v. "Nine Emperor Gods Festival."

44 Subhamoy Das, "Navadurga and the 9 Forms of the Hindu Goddess Durga," Learn Religions, August 29, 2018.

CHAPTER 4

The Triads of Angels

The angel triads, as defined by a mystic and accepted by church leaders of the time, are never discussed in the Bible are described below. These are the descriptions developed by Pseudo-Dionysius the mystic and accepted by Christians, including church clergy and theologists. Note that:

a. The descriptions do not exist anywhere in any Bible.
b. Not all theologians fully agree with these descriptions, but the discrepancies are small.

The First Angel Triad

The first order or triad of angels includes the seraphim choir, the cherubim choir, and the throne choir. These three angel choirs are pure spirits of contemplation. Their holiness is so intense that the human mind is incapable of comprehending their levels of adoration and participation in the Divinity. They are described as follows:

Seraphim

Seraphim are the highest order or choir of angels. These particular angels serve as guardians or attendants before God's throne. They praise God, calling out, "Holy, holy, holy is the Lord of hosts." *Host*, when used in this sort of context in the Bible, refers to an angel army.[1] This is part of the Sanctus prayer in Christian Eucharistic services, which is recited as follows: "Holy, holy, holy Lord, God of power and might, heaven and earth are full of Your glory. Hosanna in the highest. Blessed is he who comes in the name of the Lord. Hosanna in the highest."

The noun *hosanna*, a cry of praise, claims a long pedigree. It is believed to have held this place in the liturgy since the fifth century. Its origins in Christian worship may go back to the second century. It is derived from a biblical Hebrew phrase meaning "Pray, save us." *Hosanna* appeared in Greek and Latin before arriving in English in the earliest translations of the Bible. Any kind of thanks and adoration aimed at God could be considered a hosanna. In a church service, many of the prayers and hymns are hosannas.

The Sanctus is recited by the congregation during the Communion part of the divine Eucharistic service. The words come from Isaiah's vision of heaven (Isaiah 6) and John's vision of heaven in Revelation (Revelation 4). Further, they include a phrase from the Palm Sunday Gospel, "Blessed is he who comes in the name of the Lord. Hosanna in the highest" (Matthew 21:9).

---

1 Dolores Smith, "Who Are the Seraphim? The Seraphim in the Bible," Christianity.com, April 16, 2019.

The use of "Holy, holy, holy" together with "Blessed is he who comes in the name of the Lord" has some doctrinal implications. When they heard the song or shout of the angels, "Holy, holy, holy," Isaiah and John were said to be in the presence of God. The Palm Sunday acclamation also states, "Your Savior is here." There is a connection between the Sanctus and the doctrine of the real presence in the Lord's Supper.

"The Lord of hosts" means "the Lord of armies," as noted earlier. Angels are God's army sent to do God's bidding and deliver His messages. Repeating *holy* three times in prayer, called a Trisagion, is significant to the subject of *Angel Communication Code* and the occurrences of the number 3 in our code. It is a Holy Trinity (Father, Son, and Holy Spirit) reference. It is also a reference to beginning, middle, and end. It is repeated to highlight God's supreme holiness yet is used in this exact manner only two times in the Bible (Revelation 4:6–9 and Isaiah 6:3). The number 2 is a significant number in the ET communication code as it is the first prime number. So, the first prime number in our code is now linked to the highest order of holiness. Holiness is another concept that is one of the most difficult things in the Bible and all Christianity to grasp. In the simplest terms, the holiness of God refers to absolute moral purity of God and the unclosable moral gap between God and His human creation. The subject is way deeper and more complicated than this simple description.

There are believed to be seven seraphim angels, in order of importance as follows:

1. Michael
2. Gabriel
3. Raphael
4. Jeremiel
5. Raguel
6. Zerachiel
7. Remiel.

Only Michael and Gabriel are specifically identified by name anywhere in the Bible. In certain places in the Bible, they are said to be archangels and not seraphim, which is confusing and remains without explanation. Seraphim angels are described as glorious exotic creatures. Religious texts describe them as radiating brilliant light like flames of fire. Each seraph has six wings (three pairs), which serve different purposes:

1. Two wings cover their faces, shielding them from becoming overwhelmed by looking directly at God's glory.
2. Two wings cover their feet, symbolizing their humble respect for and submission to God.
3. Two wings enable them to fly around God's throne in heaven, representing the freedom and joy that come from worshipping God. The seraphim's bodies are covered with eyes on all sides, so they can constantly watch God in action. [2]

It is important to again point out that seraphim and cherubim are the only angel choirs the Bible referred to as having wings and using those wings to fly.

"Holy, holy, holy" is a statement and prayer message of incredible importance made by the highest order of angels, the seraphim, yet it is used in this precise way only two times in the Bible.[3] It is a grand concept

2 Whitney Hopler, "Seraphim Angels: Burning with Passion for God," Learn Religions, August 25, 2020.

3 Hope Bolinger, "Why Is God Called 'Holy, Holy, Holy'? Revelation 4:8 Explained," BibleStudy.org, July 12, 2019.

now linked to the numbers 2, 3, and 7 of our extraterrestrial communication code drivers, 2, 3, 5, and 7, with high purpose.

Cherubim

Following seraphim in the angelic hierarchy are the cherubim angels. According to Thomas Aquinas, the cherubim are characterized by knowledge, in contrast to seraphim, which are characterized by their "burning love to God."[4]

The cherubim are described as winged creatures that appear in a number of places in scripture. Cherubim are the first to appear in the Bible (after the serpent—fallen angel). In the book of Genesis, the first book of the Bible, cherubim were charged with the task of protecting the Garden of Eden after the Fall and banishment of Adam and Eve from the Garden. From Genesis 3:24: "He drove them out; and at the east end of the garden of Eden He placed the cherubim, and a sword flaming and turning every way to guard the tree of life." In the New Testament, cherubim angels are referred to as celestial attendants of the apocalypse, described in Revelation 4–6.

The Bible describes two golden figures of the cherubim with their wings stretched over the mercy seat on the Ark of the Covenant. Exodus 25:18–21 reads as follows:

> You shall make two cherubim of gold; you shall make them of hammered work, at the two ends of the mercy seat. Make one cherub at the one end and one cherub at the other; of one piece with the mercy seat you shall make the cherubim at its two ends. The cherubim shall spread out their wings above, overshadowing the mercy seat with their wings. They shall face one to another; the faces of the cherubim shall be turned toward the mercy seat. You shall put the mercy seat on the top of the ark; and in the ark you shall put the covenant that I shall give you.

God promised to meet with His people there. Exodus 25:22 reads, "There I will meet with you, and from above the mercy seat, from between the two cherubim that are on the Ark of the Covenant, I will deliver to you all My commands for the Israelites."

Many believe that cherubim are actually symbolic of redeemed humanity. According to this view, the perfections of humanity that were lost at the Fall are now reflected in the cherubim. They represent, not fallen humanity, but redeemed humanity. They are symbolic of what God has done for us. Consequently, they are placed near the symbolic presence of God. Depictions of the cherubim were woven into the veil of the tabernacle. Exodus 26:1 reads, "Moreover you shall make the tabernacle with ten curtains of fine twisted linen, and blue, purple, and crimson yarns; you shall make them with cherubim skillfully worked into them." In addition, two large cherubim were carved and set next to the Ark of the Covenant in the temple. First Kings 6:23, 27 reads: "In the inner sanctuary he made two cherubim of olivewood, each ten cubits high. … He put the cherubim in the innermost part of the house; the wings of the cherubim were spread out so that a wing of one was touching the one wall, and a wing of the other cherub was touching the other wall; their other wings toward the center of the house were touching wing to wing."

---

[4] D. Keck, *Angels and Angelology in the Middle Ages* (Ukraine: Oxford University Press, 1998), p. 25.

Not only is the exact identity (i.e., by name) of the cherubim not known, but also it is not known how they appeared. Some identify them with the living creatures of Ezekiel 1 and Revelation 4. If that is accurate, then they would have four faces. Revelation 4:7 reads, "The first living creature like a lion, the second living creature like an ox, the third living creature with a face like a human face, and the fourth living creature like a flying eagle." Conversely, these angels are said to have presented themselves to us as humanlike in appearance, double-winged guardians of God's glory.

Clearly the cherubim are important. They guarded the Garden of Eden, their form was fashioned on the Ark of the Covenant, and two large carved cherubim were placed in the temple of Jerusalem. As to their exact identity and appearance, no one knows. It is not certain that they are the creatures identified in Ezekiel 1 or Revelation 4.[5] They are certainly not at all the pudgy winged babies as depicted in paintings by Raphael during the Renaissance period. In fact, Lucifer was said to be a cherub before he became a fallen angel (Satan) and lost his holy status. Much about cherubim remains a mystery.

The following are with respect to the subject of *Angelic Communication Code*:

a. A large portion of the cherubim's responsibilities was/is to protect the Ark of the Covenant, arguably the most sacred object ever to exist in Christianity. The most important holy task is linked to the first number of the code development, prime number 2.
b. The whereabouts of the Ark of the Covenant are not known. Perhaps it is not on earth and the purpose of the code to ET communication is to lead us to it. It is not unreasonable to assume that if an artifact of such high importance was to be hidden, the people who hid it would make a map to its location. Historic information tells a different story, but the accuracy of history is not always correct. The map leading to the ark is therefore still undiscovered.

Thrones

Throne angels are more often than not referred to as ophanim. In ancient Hebraic, the word *ophanim* was thought to have meant "wheels." Thrones/ophanim are the wheels of God's throne. Other possible spellings of this word were "auphanim" or "ofanim," and also "galim."

According to the book of Ezekiel (3:15), Jehoiachin, king of Judah; a priest named Ezekiel; and ten thousand Jews were captured during an invasion of Babylonian in 597 BC and were taken to a village called Tel-abib. Five years later, when they were still in exile, God approached Ezekiel near a river in Chebar (modern-day Iraq) and commanded him to prophetic ministry service by showing him an extraordinary vision (Ezekiel 1:1–2). As part of God's great plan to call Israel to repentance, He chose to open the heavens before Ezekiel's temporal eyes. The ophanim described in Ezekiel's vision are impossible to define apart from the full scope of the revelation.

Ezekiel sees an ominous fiery lightning cloud wafting toward him from the north. Four illuminated beings blaze brightly within the cloud. Although the beings resemble the form of a man, they are far from mortal. Each has four faces—one human, one lion, one ox, and one eagle. They're completely covered with eyes from the tops of their heads to the tips of their glowing calflike feet. Their human-shaped hands tuck inside each of their four wings. One set of wings stretches outward to connect with the wings of its counterparts,

---

5 Don Stewart, "Who Are the Cherubim?," Blue Letter Bible, n.d.

while the other set shrouds the ophanim's own bodies (Ezekiel 1:4–11, 10:12). This is a very weird description, to say the least. Not how most people think of an angel's appearance. Maybe they were not angels at all.

Ezekiel does not identify these beings by name in Ezekiel 1, where the description of what he envisions begins. Later, in chapter 10, he identifies the creatures as cherubim. These angelic creatures are the same ones associated with the Ark of the Covenant images (Exodus 25:18–22) and the angels most frequently cited in the Hebrew Bible. Cherubim are known as the guardians of God's throne, which makes sense when we witness what comes next in Ezekiel's vision.

Ezekiel sees and describes four massive cherubim floating toward him in his vision. The angels move back and forth in flashes like lighting at the sole discretion of God's Spirit. But then the prophet notices a wheel beneath each cherub. He records the wheels' appearance like this: "They sparkled like topaz, and all four looked alike. Each appeared to be made like a wheel intersecting a wheel. As they moved, they would go in any one of the four directions the creatures faced; the wheels did not change direction as the creatures went. Their rims were high and awesome, and all four rims were full of eyes all around" (Ezekiel 1:16–18).

The four ophanim, guided by the same spirit as the cherubim in one symbiotic entity (Ezekiel 1:20), are thought to be the chariot of God's throne. As Ezekiel's vision progresses, the reason for this designation comes to light.

Spread above the heads of the four cherubim and the ophanim, Ezekiel sees a crystal vault that is so breathtaking that the only human word he can muster to describe it is *awesome*. Above the vault, he sees a vibrant sparkling throne, and high above the throne he sees "the appearance of the likeness of the glory of the Lord." When Ezekiel witnesses the majesty of God's glory, he immediately falls facedown and hears the thunderous voice of the Almighty instructing him in his mission to bring God's judgment to rebellious Israel (Ezekiel 1:22–28).

The Bible never specifically references the ophanim as angelic beings; however, Jewish apocalyptic writers labeled them as a class of angel and listed them in their top triad of angels, along with the seraphim and cherubim, because of the ophanim's uniqueness, their supernatural power, and their close proximity to God's throne.

Ophanim are still a feature of traditional Jewish prayers sung by congregations as part of their Shabbat morning service: "The ophanim and the holy living creatures with great uproar raise themselves up; facing the seraphim, they offer praise, saying, 'Blessed be God's glory from His place.'"

Text deciphered from the Dead Sea Scrolls (4Q405) have also supported the idea of the ophanim as angelic beings, as do late sections of the book of Enoch (61:10, 71:7), where the ophanim are described with the cherubim and seraphim as angelic beings who never sleep while guarding the throne of God.

The ophanim's or thrones' role in revealing God's glory to Ezekiel was to show him things not of this earth and gain a fresh vision of God's immense greatness. "For in Him all things were created: things in heaven and on earth, visible and invisible, whether thrones or powers or rulers or authorities; all things have been created through Him and for Him" (Colossians 1:16).

The ophanim in Ezekiel's vision reveals God's supreme reign over the entire universe. The multidirectional wheels themselves remind us that we serve a God who is omnipresent, able to be in all places at all times.[6]

The thrones are a class of angels mentioned by the apostle Paul in Colossians 1:16. This verse says, "For in Him all things were created: things in heaven and on earth, visible and invisible, whether thrones or powers or rulers or authorities; all things have been created through Him and for Him." The thrones are the angels of humility, peace, and submission. If the lower choir of angels needed to access God, they would have to do so through the thrones.

The name *throne* derives from the fact that these angels carry God's throne. The beautiful throne angels serve God in the first sphere in heaven. We can find these celestial beings in both Christianity and Judaism.

In Christianity, most information about thrones is found the New Testament. In Colossians 1:16 they are mentioned as servants of Christ. Christians see them as angels full of humility. They bring peace everywhere they are, and they are full of willingness to submit to God's will. They are classified in the lowest level of the top triad of the hierarchy. This is where material things start to take shape.

The biggest mission of the thrones is to inspire faith, and they reestablish and maintain divine justice in the higher sphere. This means that the only entities that have direct access to God are the top three choirs of angels.

Jews call thrones "ophanim," "auphanim," "erelim," or "abalim." They can also be called "bene elohim," "ishim," or "arelim." In the Kabbalah they are mentioned as the Wheels of Merkabah. In the Zohar, they are actually even higher than seraphim, the top-tier angels by Christian constructs.

To Pseudo-Dionysians, thrones live in the fourth heaven. They are the third choir of angels in the first triad in heaven. Enoch called them "fiery coals," and he described the throne angel as a wheel with eyes and wings. There are some sources that claim that the number of thrones is between four and seventy. The book of Angel Raziel mentions only seven. The sixth and seventh books of Moses mention fifteen throne angels.[7] This choir of angels is ruled by Zaphkiel, Zabkiel, Sahaquiel, or Oriphiel. It depends on the source of your information.

All this information about thrones is a mixture of information taken from religious texts by various authors. The description of thrones is a human construct and, as such, is subject to interpretation. This is important when we consider the subject of *Angel Communication Code*, being that a coded message could exist in the construct of the angel hierarchy, the behavior of angels, their purpose, and the messages they have delivered.

The Second Angel Triad

The second triad of angels includes the dominions, powers, and virtues. These angel choirs deal with earth and elements. They are considered angels of creation because their purpose is all about the ordering of the universe. They are described as the "lordships" in the *De coelesti hierarchia*, which is a Pseudo-Dionysian work on angelology. This order of angels is charged with governing and ordering the laws of the created universe.

---

6 Annette Griffin, "Who or What Are the Ophanim?," Bible Study Tools, July 19, 2021.

7 "Thrones—What Is a Throne Angel?," Guardian Angel Guide, December 16, 2016.

Dominions

Dominions are the angels of leadership. They hold the highest place among the angels of the second (middle) angel triad. *Dominion* is from the Latin *dominatio* and is translated from the Greek term *kyriotites* as "lordships." The name *dominions* is taken from Saint Paul's letter to the Corinthians. In Paul's letter to the Colossians (1:16), he speaks about how God created all things through Christ: "the visible and the invisible, whether thrones or dominions or principalities or powers; all things were created through Him and for Him." This letter refers to angelic powers: one from the first triad (thrones), one from the third triad (principalities), and two choirs of the three choirs governing the second triad of angels: dominions and powers. It is through dominions that God's power and majesty is revealed.

Dominion angels are the choir that governs humanity. They do not have direct communication with God, but they do some have direct connections with humans. Dominions are said to keep the world in proper order. They manage the duties of lower angels. These angels can also give their guidance to chosen rulers of nations. Dominions do not frequently appear in front of humans despite their direct communication with some monarchs, presidents, and emperors. These angels manage order and balance.

Dominions have precedence over the angel and archangel choirs, with authority to direct and regulate them, making known to them the commands of God. Their job is to assign duties and missions to the angels of the lower spheres and to maintain universal order. Dominions are focused on any functions relative to government of the world or of the human race. They have control over the lower angels, directing them in the duties to do the will of God.

Dominions are said to look like divinely beautiful humans with a pair of feathered wings, much like the common representation of angels, but they may be distinguished from other groups by the orbs of light fastened to the heads of their scepters or their swords.

Dominions can exist in both celestial and human realms. They are said to be heavenly governors, charged with the task of striking a balance between matter and spirit, both good and evil. Dominions also function as distributors of mercy, ready to touch the entire universe and execute God's orders. It is only with extreme rarity that these angels make themselves physically known to humans. They are known for delivering God's justice in unjust situations, showing mercy toward human beings, and helping angels in lower ranks stay organized and perform their work efficiently. They also are recognized for expressing unconditional love at the same time they execute God's justice. Dominions are mentioned, but not described, many times in the Bible, and they show us they truly are leaders and messengers of the Lord.[8]

Virtues

Virtues are known for their control of the elements. Some even refer to them as "the shining ones." In addition to being the spirits of motion, they assist in governing nature. They also assist with miracles. They are also known for their work encouraging humans to strengthen their faith in God.

Virtues or "strongholds" lie beyond the thrones. Their primary duty is to supervise the movements of the heavenly bodies to ensure that the universe remains in order. They are presented as the celestial choir virtues in the *Summa Theologica*. This is the best-known work of Saint Thomas Aquinas and is often

---

8 Noreen Bavister, "Holy Angels," *The Angels* 1 (2010), http://stmichaelthearchangel.info/pdfs/Angels-2010-1.pdf.

referred to simply as the *Summa*. The *Summa Theologica* by Saint Thomas Aquinas is a collection of all the main theological teachings of the Catholic Church up to his time. It is intended to be an instructional text for theology students, including seminarians, and also for literate laypersons. It presents the reasoning for almost all points of Christian theology. Its subjects follow the cycle of God, creation, humankind, humankind's purpose, Christ, the sacraments, and back to God. Saint Thomas Aquinas believed that the angels were not all equal in power or intellect. This is important to the subject of *Angel Communication Code* because Saint Thomas was involved with the subject of angels from his earliest years, and his writings and opinions were important to the church in those days. They are still important today.

Thomas Aquinas (1225–1274) was a scholastic theologian and doctor of the church. "Doctor" is a title given by the Catholic Church to saints recognized as having made a significant contribution to theology or doctrine through their research, study, or writing.[9]

There is an elite roster of thirty-six saints who are have demonstrated by virtue of their lives and writing accomplishments that they are blessed with extraordinary wisdom and sanctity. They are the doctors of the church as of 2023. The following is the official list of these doctors, which includes when they lived and when they were indoctrinated.[10] The list is chronological by year of indoctrination.

1. Saint Ambrose of Milan (ca. 340–397), a prominent public figure, fiercely protected the Latin Church against Arianism and paganism. He left a substantial collection of writings, of which the best known include the ethical commentary *De officiis ministrorum* (377–391) and the *Exegetical exameron* (386–390). Ambrose was serving as the Roman governor of Aemilia Liguria in Milan when he was unexpectedly made bishop of Milan in 374. He also had notable influence on Augustine of Hippo [1298].
2. Saint Augustine of Hippo (ca. 354–430), North African bishop, author of *Confessions*, *City of God*, and numerous treatises, countered heretical movements and was one of the most influential theologians of the Western church, called "Doctor of Grace" [1298].
3. Saint Jerome (ca. 343–420) translated the Old Testament from Hebrew into Latin and revised a Latin translation of the New Testament to produce the Vulgate version of the Bible. He was called the "Father of Biblical Science" [1298].
4. Saint Gregory the Great (ca. 540–604), pope, strengthened papacy and worked for clerical and monastic reform [1298].
5. Saint Athanasius (ca. 297–373), bishop of Alexandria, dominant opponent of Arians, was called the "Father of Orthodoxy" [1298].
6. Saint John Chrysostom ("Golden-Mouthed") (ca. 347–407), was archbishop of Constantinople, a homilist, a writer of scripture commentaries and letters, and patron of preachers [1568].
7. Saint Basil the Great (ca. 329–379), bishop of Caesarea in Asia Minor, refuted Arian errors; wrote treatises, homilies, and monastic rules; and was called "Father of Monasticism of the East" [1568].
8. Saint Gregory of Nazianzus (ca. 330–390), bishop of Constantinople, opponent of Arianism, wrote major theological treatises as well as letters and poetry. He was called the "Christian Demosthenes" and, in the East, "the Theologian" [1568].

---

9 Larry Rice, "Doctors of the Church?," United States Conference of Catholic Bishops, 2015, accessed October 9, 2018.

10 "Chronological List of the Doctors of the Church," US Catholics, July 28, 2008.

9. Saint Thomas Aquinas (1225–1274), an Italian Dominican, wrote systematically on philosophy, theology, and Catholic doctrine. He was a patron of Catholic schools and education and was one of the most influential theologians in the West [1568].
10. Saint Bonaventure (ca. 1217–1274), Franciscan, bishop of Albano, Italy, was a cardinal [1588].
11. Saint Anselm of Canterbury (1033–1109), archbishop, was called the "Father of Scholasticism" [1720].
12. Saint Isidore of Seville (ca. 560–636), Spanish bishop, was an encylopedist and a preeminent scholar of his day [1722].
13. Saint Peter Chrysologus (ca. 400–450), archbishop of Ravenna, Italy, was a homilist and writer who counteracted Monophysite heresy [1729].
14. Saint Leo I, the Great (ca. 400–461), pope, wrote Christological and other works against the heresies of his day [1754].
15. Saint Peter Damian (1007–1072), Italian Benedictine and cardinal, was an ecclesiastical and clerical reformer [1828].
16. Saint Bernard of Clairvaux (ca. 1090–1153), French Cistercian abbot and monastic reformer, was called "Mellifluous Doctor" [1830].
17. Saint Hilary of Poitiers (ca. 315–368), one of first Latin doctrinal writers, opposed Arianism [1851].
18. Saint Alphonsus Liguori (1696–1787), founder of Redemptorists, was a preeminent moral theologian and apologist, and a patron of confessors and moralists [1871].
19. Saint Francis de Sales (1567–1622), bishop of Geneva, was spiritual writer and a patron of Catholic writers and the Catholic press [1877].
20. Saint Cyril of Alexandria (ca. 376–444), bishop, authored doctrinal treatises against Nestorian heresy [1882].
21. Saint Cyril of Jerusalem (ca. 315–386), bishop and catechist, was a vigorous opponent of Arianism [1882].
22. Saint John Damascene (ca. 675–749), Syrian monk, was a doctrinal writer and was called "Golden Speaker" [1890].
23. Saint Bede the Venerable (ca. 673–735), English Benedictine, was called "Father of English History" [1899].
24. Saint Ephrem the Syrian (ca. 306–373) counteracted Gnosticism and Arianism with his poems, hymns, and other writings [1920].
25. Saint Peter Canisius (1521–1597), Dutch Jesuit, was a catechist and an important figure in the Counter-Reformation in Germany [1925].
26. Saint John of the Cross (1542–1591), founder of Discalced Carmelites, was called "Doctor of Mystical Theology" [1926].
27. Saint Robert Bellarmine (1542–1621), Italian Jesuit, archbishop of Capua, wrote Reformation-era doctrinal defenses, catechisms, and works on ecclesiology and church-state relations [1931].
28. Saint Albert the Great (ca. 1200–1280), German Dominican, was bishop of Regensburg, a teacher of Saint Thomas Aquinas, and a patron of scientists. He was called "Universal Doctor" and "Expert Doctor" [1932].
29. Saint Anthony of Padua (1195–1231), first theologian of Franciscans, was a preacher called "Evangelical Doctor" [1946].
30. Saint Lawrence of Brindisi (1559–1619), Italian Capuchin Franciscan, was an influential post-Reformation preacher [1959].
31. Saint Teresa of Ávila (1515–1582), Spanish Carmelite, initiated the Discalced Carmelite movement and was a prolific spiritual and mystical writer, and the first woman doctor of the church [1970].

32. Saint Catherine of Siena (ca. 1347–1380), Italian Third Order Dominican, was a mystical author and was also active in support of the Crusades and papal politics [1970].
33. Saint Thérèse of Lisieux (1873–1897), a French Carmelite, wrote a spiritual autobiography describing her "little way" of spiritual perfection [1997].
34. Saint John of Ávila (1499 or 1500 to 1569), Spanish priest, preacher, and mystic, was influential in spreading the faith in Andalusia and reforming the church in Spain [2012].
35. Saint Hildegard of Bingen (1098–1179), German Benedictine abbess and mystic, recorded her visions in writing. She also wrote lyric poems, letters of advice and prophecy, and treatises on medicine and physiology [2012].
36. Saint Gregory of Narek (950–ca. 1005), Armenian monk and poet, is also recognized as a saint in the Armenian Apostolic Church [2015].

Virtues represent unshaken fortitude in advancing the cause of God. Virtues are linked with the premise of holy grace and deliver blessings from the heavens. All of nature is subject to their control; the seasons, stars, and moon, even the sun, are subject to their command. Virtues are the angels through which God works His miracles. The name *virtue* represents a powerful, brave, and determined attitude welling forth into all the virtues' godlike strength. This deep-rooted courage is the characteristic of all who work miracles in God's name.

Virtues provide courage, grace, valor, and heroic deeds. Virtues have the task of dispensing the graces of God, which make difficult things easy. If we are to do great things for God, we learn this through virtues.[11] To live a "virtuous" life is to live a life of high moral standards.

Powers

The powers are considered warrior angels as they defend against evil, defending not only the cosmos but also humanity. They are also called the powers because they have power over the devil himself, in order to restrain the power of the demons. They also help people who are struggling with passions and vices to cast out any evil offered by the enemy. The powers or "authorities" appear to collaborate in power and authority with the principalities. The powers are the angelic representatives of all lawful authority, ecclesiastical and civil.

Powers are warrior angels against evil spirits defending the cosmos and humans. They are known as rulers. These spirits hold one of the most critical missions and are responsible for maintaining the order between heaven and earth. They are the critical line of defense and of battle during heavenly warfare with Satan. Powers maintain order around planet Earth and protect it from being overthrown by Satan. They also oversee the distribution of power among humankind.

The choir of powers directs the lower choirs on how to order creation—"and the stars of heaven shall fall, and the powers that are in heaven shall be shaken" (Mark 13:25). The powers are also tasked with guarding the celestial byways between the two realms. They are the bearers of conscience and the keepers of history and are warrior angels.

Their power is of a more intensive nature than that of the virtues, and the Devil must give way before them. It is said that the powers guard priests more carefully in their ministries. Priests under the protection of the

[11] Noreen Bavister, "Holy Angels."

powers have great influence over souls, and their work is very important. Powers are assigned to priests, who are confessors of very devout souls. These angels lead the priests to self-knowledge and instill in them a desire for perfection.[12]

The Third Triad of Angels

Principalities

The principalities, as do the choirs above them, have command over the lower angels. They also direct the fulfillment of divine orders. They are also known as princedoms or rules as they directly watch over large groups and institutions, including nations and the church. They also ensure fulfillment of the divine will. While these angels are still wise and powerful, they are in the triad farthest from God in the angelic hierarchy, so they are better able to communicate with human beings in ways we can understand.

Archangels

Archangels are called "the great heralds of the good news" because God sends them to deliver important messages to humankind. Most notable are the only two angels cited by name in the Bible: Michael and Gabriel. They are the ones who communicate and interact with us in the most profound ways. The archangel Michael is believed to be the angel who delivered God's divine inspiration to John in the book of Revelation and is known as the protector of the church, guarding it from evil.

The archangel Michael is most widely known for his role in expelling Lucifer from heaven. Gabriel is first mentioned in the book of Daniel and helps Daniel in his mission on earth. Later on, Gabriel appears to Zechariah and the blessed Virgin Mary, delivering the greatest message ever about God's intent to incarnate through the Virgin Mary and live with us through their son Jesus.

Angels

The angels of the angel choir are identified as the closest angels to the earthly world of human beings. One of the greatest characteristics of the angel choir is that they are the most caring and are directly intimate with people to assist those who pray to God for whatever reason they pray. People are more easily able to connect with these angels. Their mission is not so complex as is the mission of the higher angel choirs, but their mission is equally as, if not more, important. Closeness to humans is the mission of the angel who is part of the angel choir. Their mission is all about closeness to all human emotion, which is precisely what God desires for Himself. Circle back to the purpose for the incarnate birth of Jesus and the prior discussion of holy imperfection in *Angel Communication Code*. The word *incarnate* means "invested with flesh or bodily nature and form, especially with human nature and form," and that word is applicable in many different religions in which a god takes on an animal or a human form. In Christianity, the word *incarnation* is used in the context of "the union of divinity with humanity in Jesus Christ."[13]

It would appear that the lowest choir of all the nine angel choirs is charged with oversight of what God has always sought for Himself. This concept speaks again to the idea that nothing is perfect and God Himself may have actually had a blemish in his works. There are biblical aspects of creation over which

---

12 Ibid.

13 *Merriam-Webster Dictionary*, s.v. "Incarnate."

God did not have control in His works relative to the human equation and therefore He seeks. God has never achieved the creation or the development of a population of human beings on earth fully loyal to His commandments. His creation was and is flawed. From Adam, Eve, and their disobedience in the book of Genesis to today, there have always been issues with human loyalty and obedience to God. Loyalty and obedience to His commandments are the only things God requires, yet they are the only things He seems to be unable to receive from His created people. He destroyed most of the living things on earth in the days of Noah because of this failure.

The descriptions of the various angel choirs and their roles are not specifically defined or even outlined in any sort of detail, not even in a general way, in scripture. In scripture, we are supposed to be encouraged by the knowledge that God's angels are at work regardless of hierarchy.

Much has been written about each of the nine angel choirs since the time of the development of the angel hierarchy by Pseudo-Dionysius the mystic. It was all defined by humans and mystics. Very little, if anything, about angels in and of themselves is defined in the Bible. Whatever angels have been installed into our lives as divine messengers, and however they have been installed, their message is always, in whole or in part, about obeying the message to get to a place that is not of this earth. That is precisely what we are doing in *Angel Communication Code* by trying to figure out how to establish two-way communication with our ET brothers and sisters.

The less conspicuous message from angels is that the prime numbers 2, 3, 5, and 7 that they put before us must be put together and analyzed to find the key to ET communication.

CHAPTER 5

Word of Angels

Angel Communication Code considers as relevant to the subject hereof any messages in the Holy Bible as either the word of angels or the Word of God as written by humans. The holy words of angels as cited in the Bible are one thing, and the designed words and structure of angels devised by humans is another. Consistent with the guiding principles of cryptography, it is logical to expect that the best way for a cryptologist to design and deposit a coded message is to deposit that message in the most accessible possible place for its intended recipients to find it. How does that premise apply the word of angels?

The bestselling and most widely distributed book of all time through the year 2023 is the Holy Bible. The Bible is where the existence of, actions of, and messages from angels are mostly documented.

The Bible is the Holy Scripture of the Christian religion. It claims to tell the true history of the earth from its earliest creation on through the spread of Christianity to the prophesied end of times. The Old Testament and the New Testament have both undergone changes over the centuries. This includes the publication of the King James Bible in 1611 and the addition of several books that were subsequently discovered.

The Old Testament is the first section of the Bible, covering the period from the creation of Earth through Noah and the Flood to Moses, and closing with the Jews' exile to Babylon. The Old Testament is very similar to the Hebrew Bible, which has origins in the ancient religion of Judaism. The exact beginnings of the Jewish religion are unknown. The first currently known mention of Israel is an Egyptian inscription from the thirteenth century BC.

The earliest-known mention of the Hebrew God, Yahweh, is in an inscription relating to the king of Moab in the ninth century BC. It is speculated that Yahweh was possibly adapted from the mountain god Yhwh in ancient Seir or Edom.

It was during the reign of Hezekiah of Judah in the eighth century BC that historians believe the Old Testament began to come together. It was the product of scribes recording royal history and heroic legends. The final form of the Hebrew Bible developed over the next two hundred years when Judah was overtaken by the aggressively expanding Persian Empire.

The Hebrew Bible was translated into Greek in the third century BC following the conquest of Alexander the Great. The Greek translation was called the Septuagint (previously cited) and was to be included in the Great Library of Alexandria. The Septuagint was the version of the Bible used by early Christians in Rome.

The New Testament tells the story of the life of Jesus and the early days of Christianity, most notably, Paul's efforts to spread Jesus's teaching. It holds twenty-seven books, all originally written in Greek. The sections of the New Testament concerning Jesus are called the Gospels, which were composed approximately forty years after the earliest written Christian texts. The Epistles are the letters of Paul. Paul's letters were distributed by churches sometime around AD 50, just before Paul's death. As circulation of these letters continued, they were eventually formed into books.

Some in the church, inspired by Paul, began to write and circulate their own letters, so historians believe that some books of the New Testament attributed to Paul were in fact written by disciples and imitators.

As Paul's words were circulated, an oral tradition began in churches telling stories about Jesus, including teachings and accounts of postresurrection appearances. Sections of the New Testament attributed to Paul talk about Jesus with a firsthand feeling, but Paul never knew Jesus except in visions he had, and the Gospels were not yet written at the time of Paul's letters.[1]

The Bible is preached in church as the incontestable Word of God written for His people. The earliest King James Version of the Bible contained the so-called books of the Apocrypha for a total of eighty books, instead of the sixty-six of today's Bible.[2] For various reasons, these extra books were deemed not to have been inspired by the Holy Spirit and were removed from later editions. The Protestant (King James Version [KJV]) Bible currently consists of sixty-six books, which are considered divinely inspired. There are thirty-nine books contained within the Old Testament and twenty-seven books in the New Testament. The books of the Catholic Bible, however, include all sixty-six KJV books plus seven extra books. The additional books in the Catholic Bible are found in the Old Testament and include Tobias, Judith, Wisdom, Ecclesiasticus/Sirach, Baruch, 1 Maccabees, and 2 Maccabees, bringing the total to seventy-three books.

The Holy Bible, inclusive of all versions, is the bestselling book of all time with more than fifty billion copies sold and distributed. The entire Bible has been translated into 683 languages. There is a dramatic difference between the Holy Bible and other collections of holy texts.

The oldest known Bible is the Codex Vaticanus, so called because it is the most famous manuscript in the possession of the Vatican library. It is believed to be from the fourth century and is thought to be the oldest (nearly) complete copy of the Greek Bible in existence. It is missing most of the book of Genesis, the book of Hebrews from 9:14 to the end, the Pastoral Epistles, and the book of Revelation. These parts were lost by damage to the front and back of the volume, which is common in ancient manuscripts. The writing of that text is in capital letters (called *uncial* script) without spaces between words (*scriptio continuo*). It is arranged in three columns on each page. Like other early manuscripts, its text is somewhat shorter than the later manuscripts and less harmonistic in parallel passages of the Synoptic ("presenting or taking the same or common view") Gospels. It often agrees with the texts presumed to underlie the ancient Coptic, Syriac, and Latin versions against the later Greek manuscripts. It is relatively free of obvious transcriptional errors and is usually taken as the best representative of the ancient Alexandrian form of the New Testament text.

For many years, the Codex Vaticanus was highly valued by scholars who knew next to nothing about it other than its antiquity. Johann Bengel, in his *Gnomon of the New Testament*, makes the following remark concerning it:

---

1 "The Bible," History.com, April 23, 2019.

2 Jack Kelly, "80 Books or 66?," Grace Through Faith, February 4, 2007.

"The number of witnesses who support each reading of every passage ought to be carefully examined: and to that end, in so doing, we should separate those codices … which are known to have been carefully collated, as, for instance, the Alexandrine, from those which are not known to have been carefully collated, or which are known to have been carelessly collated, as for instance the Vatican MS [manuscript], which otherwise would be almost without an equal."

The Codex Vaticanus was highly coveted by the librarians of the Vatican and yet was virtually inaccessible to competent scholars for most of the nineteenth century, during which time its reputation was magnified by its mystique. When at last it was fully revealed, in many places it was seconded by the later discovered Codex Sinaiticus, setting the stage for the influential text of Tischendorf 1869 and the even more important edition of Westcott and Hort 1881.

Tischendorf 1867 was the first reliable edition of the manuscript to be published. It was preceded by the manuscript of Cardinal Mai, and then by Hansell 1864. Mai's edition was soon discovered to be very faulty, and scholars generally did not accept it as being more reliable than the collations that were already available to them. Hansell's edition was based upon Mai's.[3]

The *Garima Gospels* is the world's earliest illustrated Christian book. It was saved by a British charity, which located it at a remote Ethiopian monastery. These Garima Gospels are named after a monk who arrived in the African country in the fifth century. Legend says he copied them in just one day. It is magnificently illustrated (below) and the colors are still vivid, thanks to the Ethiopian Heritage Fund.

3 "Codex Vaticanus," Bible-Researcher.com.

Abba Garima arrived from Constantinople in AD 494, and legend has it that he was able to copy the Gospels in a day because God delayed the sun from setting. The incredible relic has been kept ever since in the Garima Monastery near Adwa in the north of the country, which is in the Tigray region at seven thousand feet. Experts believe it is also the earliest example of bookbinding still attached to the original pages. The survival of the Gospels is incredible considering the country has been under Muslim invasion; an Italian invasion and a fire in the 1930s destroyed the monastery's church. These Gospels were written on goatskin in the early Ethiopian language of Ge'ez. [4]

C. S. Lewis was not only the extraordinarily gifted writer of now-classic works such as *Mere Christianity* but also one of the greatest medieval English literature scholars of his time, who served at both Oxford and Cambridge Universities. He wrote in one of his essays that the Bible is different from every other "holy" book in the world. He further wrote that other sacred texts come across more like mythology. In fact, *The Chronicles of Narnia* came out of both Lewis's command of mythology and his Christian faith.

Scottish Bible scholar John Murray, of Westminster Theological Seminary, began his essay on the subject of who wrote the Bible with a simple but carefully crafted summary: "Christians of varied and diverse theological standpoints aver that the Bible is the Word of God, that it is inspired by the Holy Spirit, and that it occupies a unique place as the norm of Christian faith and life." But as the Lord God ordains whatsoever comes to pass, He does so by means: secondary forces orchestrated by God to bring about His will. Thus it is with the Bible.

The Holy Bible is assembled to include the interpretations of forty independent authors, possibly fewer or possibly more, writing across at least fifteen hundred years. The accepted list of books in the Bible is called the canon. The Catholic Bible Old Testament canon is all about history. In the days of Jesus, however, there was no official Old Testament canon. In those days, there were several different collections or versions of scripture in circulation among the Jews. The two most prominent and accepted versions were:

1. the Septuagint, an early Greek translation of the Old Testament holding forty-six books and
2. the Old Testament in Hebrew, containing thirty-nine books and excluding the books of Tobit, Judith, Wisdom, Sirach (Ecclesiasticus), Baruch, and 1 and 2 Maccabees. It also excludes chapters 10–16 of Esther and three sections of Daniel: Daniel 3:24–90, Daniel 13, and Daniel 14. These excluded books and chapters are called the deuterocanonical book, which means the "second canon."

Hebrew-speaking Jews during the time of Jesus would have used the Hebrew Old Testament; the Greek-speaking world around them used the Greek Septuagint. The authors of the New Testament books also directly referenced the Septuagint most of the time. This was the version most commonly used in the early days of the church. Because the Septuagint was the version most used and accepted in the church's earliest days, the Catholic Church uses the Septuagint's canon of Old Testament books in the Roman Catholic Bible. The list of the Old Testament books of the Catholic Bible is firmly rooted in history.

The New Testament is an entirely different matter with respect to how it came to be. It was not a question of what ancient texts of Jewish scripture should or should not be in the canon. It was a matter of what new

4 "Ethiopian Bible Is the Oldest and Most Complete on Earth," OrthoChristian.com, June 2016.

books about Jesus and the Christian life were true and could be accepted as inspired texts of Christianity. The question was different, but the process to get there was quite similar.

All the New Testament books of the Catholic Bible were selected because the church's bishops agreed that those books alone were divinely inspired. They were considered accurate teachers of the true faith received from Jesus and the apostles. It took a few hundred years to complete this process of officially defining the Christian canon of both the Old and New Testament.[5] That is the short version of how the Catholic Bible came to be from Genesis to Revelation and everything in between.

The King James Version (KJV) was the result of the Protestant Reformation of the sixteenth century. Protestant reformers of the time did not accept certain books of the Catholic Bible. It was the Protestant Reformation that led to the development of the King James Version of the Bible. Protestantism is a Christian religious movement that came out of the Reformation and sowed its seeds in Northern Europe during the early sixteenth century. It was the reaction to the harsh medieval Roman Catholic doctrines and practices of those days. Protestantism grew to become one of the top three most powerful denominations in Christianity, rivaling Roman Catholicism and Eastern Orthodoxy. After several European religious wars in the sixteenth, seventeenth, and nineteenth centuries, Protestantism spread throughout the world. Wherever Protestantism went, the result was an influence on the local social, economic, political, and cultural life of the region.

In the sixteenth century, the word *Protestant* referred to the two schools of thought that arose in the Reformation. One was the Lutheran, and the other was the Reformed. In England in the early seventeenth century, the word was used to identify Orthodox Protestants, as opposed to those who were regarded by Anglicans as unorthodox, as were the Baptists or the Quakers. Roman Catholics, however, used it for all who claimed to be Christian but opposed Catholicism. Roman Catholics included Baptists, Quakers, and Catholic-minded Anglicans. The English Toleration Act of 1689 was "an Act for exempting Their Majesties' Protestant subjects dissenting from the Church of England." The act only allowed for the toleration of the opinions known in England as "orthodox dissent." Protestantism is forever linked to the sixteenth-century Reformation.[6]

King James I of England was born in 1566. Then 451 years later, in 1611, the New King James Version of the Catholic Bible was published. The King James Bible is one of the most printed books ever. King James came to the throne during very confrontational and confusing religious times. His half-sister, Queen Mary I, a.k.a. "Bloody Mary," a Catholic, executed nearly two hundred fifty Protestants during her short reign. This was only about fifty years before James came to power. Elizabeth, as queen, affirmed the legitimacy of her father Henry VIII's Anglican Church but maintained an agreement where Protestants and Puritans were allowed to practice their own varieties of the religion. The Anglican Church was under attack from Puritans and Calvinists seeking to extinguish bishops and their hierarchy. In the 1640s, these disputes would become catalysts for the English Civil War. During James's reign, however, the conflicts were expressed in a very different forum, which was translation.

The translations of ancient texts exploded in the fifteenth century. Scholars in Italy, Holland, and elsewhere perfected the Latin of Cicero and learned Greek and Hebrew. The resurgence of these languages combined with advances in printing capabilities allowed access to knowledge of not only the secular (the pagan

[5] "Books of the Catholic Bible: The Complete Scriptures," BeginningCatholic.com.

[6] *Encyclopedia Britannica*, s.v. "Protestantism, Christianity," Martin E. Marty et al., April 4, 2023.

classics) but also the sacred (the Bible in its original languages). The new translated texts were in high demand from people capable of reading the ancient languages. Protestant scholars used their new learning to render the Bible into common language intended to provide people a more direct understanding of, and relationship with, God. In England, the publication of translations began with William Tyndale's 1526 Bible. The translated texts evolved into the Geneva Bible, which was completed by the Calvinists, whom Queen Mary had exiled to Switzerland.

This Geneva Bible was the Bible most popular among the reformers at the time of James's ascension to the throne. Its circulation was a serious threat to Anglican bishops. Not only did the Geneva Bible begin to replace their translation (the so-called Bishops' Bible), but also it appeared to challenge the supremacy of secular rulers and the bishops' authority. One of its most critical descriptions compared the locusts of the apocalypse to swarming hordes of "prelates" dominating the church. Others referred to the apostles and Christ Himself as "holy fools." This was a term intended to instigate a challenge to "all outward pomp" and the decadence of the Anglican and Catholic Churches.

In 1604, King James, who was also a religious scholar, sought to unite these factions and his people by way of creating a single, universally accepted Bible. His idea was proposed at a conference of scholars at Hampton Court by a Puritan, John Rainolds, who was the seventh president of Corpus Christi College. Rainolds hoped that James would reject the Bishops' Bible, but his plan failed when the king insisted that the new translation be based on it and also condemned the "partial, untrue, seditious" notes of the Geneva translation. The result was the King James Version of the Catholic Bible, which contains sixty-six books, versus seventy-three (seven fewer than the Catholic version).[7]

The singular message, the "scarlet thread" of truth that binds each of the historic versions of the Bible books, is the witness of Jesus of Nazareth and the witness of the Holy Spirit. They all converge to make the Bible, alone against all other revered texts, a revelation of God to humankind. The Bible self-identifies as a supernatural "word from another world." God could have spoken directly or "immediately," but He (mostly) chose to do so through His agents, the angels and His only begotten Son. The Bible book writers were prophets, priests, kings, servants, lawyers, anglers, scholars, and the uneducated who wrote under the inspiration of the Holy Spirit.[8]

The Bible is where Christians and even non-Christians look for hope, salvation, and life-changing answers. The Bible is the book where angels are the messengers of God and are extremely important for the delivery of divine messages to humanity. *Angel Communication Code* points to the angels for another message that reveal the messages and gives clues for ET communication. Those angelic messages are in the form of recurring numbers (2, 3, 5, and 7) that are presented during critical events in critical places and at critical times within all versions of the Bible. It is all about the humans of Earth communicating with beings not of this earth, whether they are spiritual or biological—or perhaps both.

The number of Bibles in the world that were available to people before the invention of the printing press by Gutenberg in the mid-fourteenth century AD was quite small. It was still the most common book in the Christian world before that time. The enormous expenditure of time and money required to reproduce such a complex book by hand meant that hand-copied reproductions of the Bible were extremely rare and very valuable—not available or affordable to most common folk.

---

7 Joel L. Levy, "How the King James Bible Came to Be," Center for Jewish History, June 19, 2017.

8 Michael A. Milton, "Who Wrote the Bible?," Biblestudytools.com, May 2, 2022.

The problem of the Bible's availability and affordability began to change with the development of printing technology in the mid-1400s. In the years that followed, the first Gutenberg press produced its first Bible in 1454. It has been estimated that European printers produced fifteen million books. The majority of those books were Bibles. As technology improved and literacy rates increased, the Bible production rate increased. By 1815, the total number of Bibles printed was around 1.3 billion.

A key milestone in the history of Bible distribution came in the early nineteenth century when the first Bible Societies were founded. These organizations ordered Bibles in vast numbers, but gave them away or sold them at cost as part of their Christian mission. From this point on, the majority of Bibles were freely distributed rather than sold. A study produced by the British and Foreign Bible Society (founded in 1804) calculated that approximately two and a half billion Bibles were printed between 1815 and 1975. Today, Bibles are printed at a rate of approximately eighty million per year.[9] That is many Bibles, and the Bible is the Word of God as authored by mortal men. The Bible is the place where the messages of angels and their encounters with humans are documented for our interpretation. This is clearly a good place to look closer for angelic clues to an ET communication code key.

Angels are woven into the Bible story, which covers thousands of years of human history in the world of God's people. It has been edited, supplemented, translated from numerous languages, and copied many times. In addition, there are approximately forty-five thousand denominations of Christianity in the world today. It is not surprising that certain details of the story, like any other story over time, have changed. What is truly amazing is that the fundamental messages delivered to us through the Bible have not changed. Messages delivered to us from God Himself via the Bible authors, including human-authored angel messages, remain fundamentally intact from the original books of the Bible.

Contrary to popular belief, there are no lost books of the Bible and no books missing from the current version of the Bible. Every book that the Christian leadership intended to be in scripture is there.

Throughout the history of the Christian church, there have been disagreements about what books belong in the canon of scripture. The church has always agreed that the canon was closed with the apostle John's death. The Bible, as it is and always has been, is accepted as the inspired, inerrant, infallible, sufficient, and clear Word of God. There are many legends and rumors of lost books of scripture, but those books were not lost; rather, they were rejected.

The so-called lost books refer to a collection of writings in the twelfth and thirteenth centuries in Latin and published as *The Lost Books of the Bible* and *The Forgotten Books of Eden* in the 1920s.

The Old Testament section of the lost books includes eight books: the Conflict of Adam and Eve with Satan, the Secrets of Enoch (also often called Second Enoch), Psalms of Solomon, the Odes of Solomon, the Letter of Aristeas, the Fourth Book of Maccabees, the Story of Ahikar, and the Testaments of the Twelve Patriarchs. The modern translation of some books came from Ethiopia. The New Testament section includes a wide variety of writings ranging from the Infancy Gospel of Thomas (late second century) to the Epistle of Clement (a first-century church father) to the Apostles' Creed, also including writings of the early church fathers and late works that were falsely attributed to earlier writers (such as the Lost Gospel of Peter).

---

9 *Guinness Book of World Records*, s.v. "Best-Selling Books."

These books were not originally written with the intent to become part of scripture, nor were these works "lost." They were known but not accepted as the historical writings of scripture.[10] The original Bible message remains largely intact, as do the messages of the angels, be they the direct and obvious messages or the less obvious coded messages. While it is true that the Bible was the Word of God as written by human beings, it is not necessarily the only message woven into the structure of the angel hierarchy, the appearance and purpose of angels, or messages delivered to us by angels.

There is the fundamental purpose and story of the Bible, and then there may very well be a secondary purpose. The authors of the Bible books, or the Bible translators, interpreters, and editors, may very well have been masking certain clues we need to put together in order to receive and interpret a secondary message. The fundamental story and message delivered in the Bible is all about how humans need to live such that after they die, they can be accepted to eternal life in heaven. The secondary message may very well be about what we need to do to open communication with our brothers and sisters living elsewhere in the universe. Understanding both messages is necessary for our long-term survival as a species on earth and then as eternal spirits in heaven.

The New Testament cites two accounts of the genealogy of Jesus. The first is from the first book of the New Testament's Gospel of Matthew. The second version is found in the third book, the Gospel of Luke. Matthew starts with Abraham, whereas Luke begins with God. The generations referenced are identical between Abraham and David. After that, they are very different. The two accounts also disagree on who Joseph's father was. Matthew says it was Jacob, whereas Luke says it was Heli.

Traditional theologians have put forward various theories that seek to explain why the lineages are so different. Matthew's version looks like this:

1. Abraham	15. Solomon	29. Shealtiel
2. Isaac	16. Rehoboam	30. Zerubbabel
3. Jacob	17. Abijah	31. Abiud
4. Judah **and** *Tamar*	18. Asa	32. Eliakim
5. Perez	19. Jehoshaphat	33. Azor
6. Hezron	20. Jehoram	34. Zadok
7. Ram	21. Uzziah	35. Achim
8. Amminadab	22. Jotham	36. Eliud
9. Nahshon	23. Ahaz	37. Eleazar
10. Salmon **and** *Rahab*	24. Hezekiah	38. Matthan
11. Boaz **and** *Ruth*	25. Manasseh	39. Jacob
12. Obed	26. Amon	40. Joseph
13. Jesse	27. Josiah	41. Jesus
14. David **and** *Bathsheba*	28. Jeconiah	

[10] Dave Jenkins, "What Are the Lost Books of the Bible?," Christianity.com, December 17, 2020.

Matthew's version is presented in three sets of fourteen generations. The Bible says fourteen, but when you do a head count in a list, you find that it is actually one name short: $3 \times 14 = 42$. Each set is of a clear character:

- The first is thick with annotations, including four mothers and mentioning the brothers of Judah and the brother of Perez.
- The second spans the Davidic royal line but omits several generations, ending with Jeconiah and his brothers at the time of the exile to Babylon.
- The last, which appears to span only thirteen generations, connects Joseph to Zerubbabel through a series of otherwise unknown names that are suspiciously few for such a long period of time.

It is curious why Matthew cites three sets of fourteen generations yet is actually one name short. The total of forty-two generations as cited in the Gospel is achieved only by omitting several names, making the choice of three sets of fourteen appear quite deliberate. In addition, it is curious why Matthew cites in his Gospel that the genealogy of Jesus runs from a person (Abraham) to a person (David) to a period in time (the Babylonian exile), then forward to a person (Joseph). Why reference a period in time and not another person? A period in history is not a genealogical link. Theologians have offered various explanations, one of which is that 14 is 2×7. The number 7 symbolizes perfection and covenant, and is the gematria of the name David. Gematria is the practice of assigning a numerical value to a name, word, or phrase according to an alphanumerical cipher. Gematria is still widely used in Jewish culture to this day.[11]

Now let us look at Luke's version:

1. God	15. Sala	29. Aminadab	43. Judah	57. Zorobabel	71. Jannai
2. Adam	16. Heber	30. Naasson	44. Simeon	58. Rhesa	72. Melchi
3. Seth	17. Phalec	31. Salmon	45. Levi	59. Joannan	73. Levi
4. Enos	18. Ragau	32. Boaz	46. Matthat	60. Juda	74. Matthat
5. Cainan	19. Saruch	33. Obed	47. Jorim	61. Joseph	75. Heli
6. Maleleel	20. Nachor	34. Jesse	48. Eliezer	62. Semei	76. Joseph
7. Jared	21. Thara	35. David	49. Jose	63. Mattathias	77. Jesus
8. Enoch	22. Abraham	36. Nathan	50. Er	64. Maath	
9. Mathusala	23. Isaac	37. Mattatha	51. Elmodam	65. Nagge	
10. Lamech	24. Jacob	38. Menan	52. Cosam	66. Esli	
11. Noah	25. Judah	39. Melea	53. Addi	67. Naum	
12. Shem	26. Phares	40. Eliakim	54. Melchi	68. Amos	
13. Arphaxad	27. Esrom	41. Jonam	55. Neri	69. Mattathias	
14. Cainan	28. Aram	42. Joseph	56. Salathiel	70. Joseph	

Obviously, much of Luke's additional length is because his genealogy goes all the way back to Adam the son of God. This accounts for twenty-one extra names, but still leaves Luke with fifty-four entries to Matthew's forty-one. Notice that the total is seventy-seven generations, again pointing us to focus on the number

---

11 Wikipedia, s.v. "Gematria."

7, which is a critical ET communication code number with respect to the subject of *Angel Communication Code*. A side-by-side comparison of Matthew's version and Luke's version looks like this:

Matthew	Luke
	God, Adam, Seth, Enos, Cainan, Maleleel, Jared, Enoch, Mathusala, Lamech, Noah, Shem, Arphaxad, Cainan, Sala, Heber, Phalec, Ragau, Saruch, Nachor, Thara,
Abraham, Isaac, Jacob, Judah, Perez, Hezron, Ram, Amminadab, Nahshon, Salmon, Boaz, Obed, Jesse, David,	Abraham, Isaac, Jacob, Juda, Phares, Esrom, Aram, Aminadab, Naasson, Salmon, Boaz, Obed, Jesse, David,
Solomon, Rehoboam, Abijah, Asa, Jehoshaphat, Jehoram, Uzziah, Jotham, Ahaz, Hezekiah, Manasseh, Amon, Josiah, Jeconiah,	Nathan, Mattatha, Menan, Melea, Eliakim, Jonam, Joseph, Judah, Simeon, Levi, Matthat, Jorim, Eliezer, Jose, Er, Elmodam, Cosam, Addi, Melchi, Neri,
Shealtiel, Zerubbabel,	Salathiel, Zorobabel,
Abiud, Eliakim, Azor, Zadok, Achim, Eliud, Eleazar, Matthan, Jacob,	Rhesa, Joannan, Juda, Joseph, Semei, Mattathias, Maath, Nagge, Esli, Naum, Amos, Mattathias, Joseph, Jannai, Melchi, Levi, Matthat, Heli,
Joseph, Jesus	Joseph, Jesus

Both lists are coincident from the time of Abraham to the time of David. After that, they do not agree until Shealtiel (Salathiel) and Zerubbabel (Zorobabel), and then they are disharmonious again until Joseph and Jesus.

It is important to note that Jesus is not the blood of Joseph and therefore is not the blood of Abraham. Jesus is the blood of Mary and God (God does not have blood). These are all important matters to ponder. The writers of the Bible did not use the bloodline of Mary, because then the trail would not lead back to David or Abraham. There are those who make abstract arguments that allegedly link Mary to the tribe of David, but the arguments are weak. Why was it so important to link Jesus to Abraham, David, and a period in the history of God's people? Again, this shows how recorded history tells us much about the authors and sometimes less about the subject and its truth.

It is curious that in the last book of the New Testament, Revelation (22:16), Jesus says, "I am the root and the descendant of David, the bright morning star." Jesus Himself refers specifically and exclusively to King David as the point of reference for His own genealogy. David's bloodline is to Joseph and not Mary. Jesus was not a blood descendant of David. It is very curious that Jesus Himself claims this genealogical link to David. Why would the author of Revelation proclaim that Jesus said that He was the descendant of David when biologically He was not? *Angel Communication Code* contends that it is all about the numbers. Recall again here that 2, 3, 5, and 7 are the key numbers to the ET communication code. It is important to keep all the links and references to the numbers 2, 3, 5, and 7 in your mind as we move through *Angel Communication Code*. The key to the lineage is the importance of Abraham, David, and the Babylonian exile in the history of the Bible with the angel messages provided therein. The reference points of the lineage cited by Matthew are:

Abraham to David → David to the Exile → the Exile to Jesus

Let us take a closer look at Abraham, King David, and the Babylonian exile on a timescale and in terms of the specific importance of each as an historical event as it relates to *Angel Communication Code*. To go down this path, it is important to understand our frame-of-reference recording of time itself.

The Hebrew calendar is based on the anno mundi ("in the year of the world") premise. Anno mundi dates events from the beginning of the creation of the earth, those dates calculated as best as possible through scripture. Ancient civilizations designed their calendars based on the reign of kings or the cycles of the seasons as set by their various gods. In Mesopotamia one would have dated an event as "five years from the reign of King Shulgi." In Egypt, it would have been "three years after the last Opet Festival of Ramesses, who was the second of that name" or perhaps "In the tenth year of the reign of Ramesses, who triumphed at Kadesh." This method of dating was continued by the Romans, who counted their years according to three different systems in different eras: from the founding of Rome, and by emperors who ruled at a point in time.

It was Julius Caesar who reformed the calendar and renamed the months during his reign from 49 BCE (before the Common Era) to 44 BCE (or BC [before Christ]). This calendar remained in use, with periodic revisions, until 1582 CE (the Common Era). CE is the same as AD (anno Domini), the latter of which means "in the year of the Lord" in Latin. In 1582 CE, Pope Gregory XIII instituted the Gregorian calendar, which is still found in use in the present day. Christians used the anno mundi calendar and the Roman calendar in the early years of the faith. In 525 CE, a new concept for dating was introduced by a Christian monk named Dionysius Exiguus that served as the foundation for the move to the BC/AD system for tracking years.

Dionysius Exiguus invented the concept of anno Domini time in an effort to fix the date of the celebration of Easter. While he was working on this problem, Christians of the church of Alexandria were dating events from the beginning of the reign of the Roman emperor Diocletian (284 CE), who, ironically, often persecuted members of the new Christian faith. Dionysius Exiguus's goal was to bring the Eastern and Western churches into agreement as to a single day on which all Christians would celebrate Easter—the celebration of Jesus's resurrection into heaven and the most important event in all of Christianity.

This goal had been decided upon by Constantine the Great at the Council of Nicaea in 325 CE but had not yet been put into use. Dionysius Exiguus ultimately succeeded in changing the system of dating years from the Roman system and the Alexandrian system to his own new system in which his present Christian era dated from the birth of Jesus of Nazareth. This choice also eliminated another problem, which was dating historical events based on the reign of an emperor who had executed so many Christians.[12]

The only problem with this dating system was that no one actually knew exactly when Jesus of Nazareth was born. Dionysius Exiguus himself did not know when Jesus was born, and his system makes no claims to have dated that event definitively. He seems to have arrived at his calculations through a reliance on scripture and the known history of the time. This was the database for his creation of a Christian calendar, which would be acceptable to both the Western and Eastern churches of the time for the celebration of Easter.

---

[12] Robert Coolman and Owen Jarus, "Keeping Time: The Origin of BC and AD," LiveScience.com, January 14, 2022.

Dionysius Exiguus never made the claim that he knew the date of Jesus's birth, nor did he begin his quest to reform the calendar by accurately dating the birth of Jesus of Nazareth. He did it in accordance with the wishes of the pope of the time, who wanted Constantine's vision realized. The Easter celebration of the resurrection was considered the most important celebration of the church. Constantine, and those in power who followed him, wanted Easter observed by all churches on the same day. It was Dionysius Exiguus's job to make that happen. He tried to do this by making a new calendar system, which involved estimating the date of Jesus's birth.

The birth of Jesus is obviously a huge event in Christianity, but the Bible does not specifically identify the year when Jesus was born. It does, however, provide sufficient information to deduce a narrow range of dates. There is no definitive proof or agreement among Christian theologians on this matter. There are generally five recognized calculations that tackle this question, producing similar results.

For the purposes of *Angel Communication Code*, we are going with the range of between 6 BC and 4 BC as the best educated and defensible guess as to the year of Jesus's birth. In a similar way, His death on the cross is typically dated between AD 30 and 34. This is yet another case where the story of history differs depending on a given author's perception and purpose. The small range in birthdates is not important in our analysis, which covers thousands of years. What makes the vagueness of the birthdate of Jesus so strange is that it was such a big event. The whole story of the star of Bethlehem and the Three Kings is huge, and modern science has been trying to pin this date down for quite a long time. Still there is no definitive and incontestable answer to the question.

Now let us get back to our examination of possible reasons (relevant to the subject of *Angel Communication Code*) why the book of Matthew (1:1) runs through the entire genealogy leading to the birth of Jesus Christ using only three control points: Abraham, David, and the exile in Babylon.

Regarding Abraham (Abram)

Why is Abraham so important in all of this? Because he is considered the father of the world's three great monotheistic religions, Judaism, Christianity, and Islam. Abraham was born Abram in 2167 BC. God gave him the name of "Abraham" in 2068 BC when he was ninety-nine years old (Genesis 17:1).

Abraham is very much linked to our prime number communication code when we look at the numbers woven through his story. The story of the twelve tribes of Israel starts with Abraham, the first leg of the genealogical trilogy reference cited in Matthew 1:1.

Abraham was the first Hebrew patriarch. He is revered in Judaism, Christianity, and Islam. The Bible tells us that he was called by God to journey to a new land, where he founded a new nation.

As a child growing up, he lived in Ur of the Chaldeans. Scholars believe it is the place called Tall al-Muqayyar, which is approximately two hundred miles southeast of Baghdad. Later in life, Abraham settled near Hebron in Canaan.

Abram arrived in Canaan and had no children at age seventy-five with his wife, Sarah, who was not biologically able to have children. God, however, promised that Abram's children and their descendants would inherit the land. Abram was ninety-nine years of age when the Lord appeared to him and ratified their covenant, changing Abram's name to Abraham and that of his wife from Sari to Sarah.[13]

With his wife's approval and indeed her encouragement, Abraham sired a son by Sarah's maidservant Hagar. They named their child Ishmael. When Abraham was one hundred years old, God blessed him and his wife, Sarah, with their biological son, whom they named Isaac. This is a very complex dynamic on so many levels in the context of what is holy and what is not. God has rules for humans about fidelity between a husband and his wife, but in the case of Abraham, his wife, and the maidservant, He made an exception. These sorts of things can make understanding God's Word difficult at times.

Abraham is known for his intense and unwavering faith in God. In Genesis, God commands him to sacrifice his son Isaac as a test of faith. Abraham is confused by this directive, but he obeys. While Abraham is preparing to sacrifice Isaac, God provides him with a sacrificial lamb as a substitute just in time, right as the blade was coming down on the boy. That is the very short story. But think about this:

1. Abraham and Sarah were well into their old age and not able to have children, yet God made it happen.
2. God promised that Abraham would have a child with Sarah and that their descendants would be the leaders of nations, and it happened.
3. Then God flipped on them and directed Abraham to sacrifice his son to demonstrate his faith in God. Imagine Abraham's confusion and frustration when he received this directive from God.
4. True to his faith, Abraham made preparations to lay his son down on an altar and slit his throat to prove his faith to God.
5. God stepped in at the last second as the blade was coming down and stopped Abraham. He provided a lamb for the sacrifice and saved the boy.
6. Isaac grew up and begot two sons he named Esau and Jacob. Jacob had twelve sons, who became the patriarchs of the twelve tribes of the nation of Israel.

It can be argued that this was how God first came to understand something about humans he did not know before. God is not human. Perhaps He did not understand why He promised eternal life for those who were faithful to Him, and yet the people He created were, more often than not, prone to unfaithfulness and sin. A truly faithful man can love and trust God so much that he is willing to kill his only son as a sacrifice. He can do this trusting that God will fulfill His promise to make that sacrificial son the ruler of nations. This in turns means that Abraham trusted that God would have to resurrect his son from the dead to fulfill that promise if Abraham had followed through and sacrificed the boy. This is perhaps how God came to the decision to create His own earthly begotten Son, have Him sacrificed on the cross, and then resurrect Him with the function of forgiving human sin. This would surely get the people to obey the Word of God.

Abraham worshipped his God exclusively as "God Most High." He is often considered the first monotheist.[14]

Two other significant aspects of Abraham's story are specifically relevant to *Angel Communication Code*. The first is that Abraham was physically visited by three angels that delivered the messages about Abraham's children and the destruction of Sodom and Gomorrah. Why three angels? Why not one angel? Note that

---

13 "Who Was Abraham?," Bibleworld, February 3, 2022.

14 *Encyclopedia Britannica*, s.v. "Abraham, the First Hebrew Patriarch.

some accounts of this story say one of the three was actually God. The important clue in these events is the inclusion of the number 3.

The second significant aspect is found in the book of Abraham, which is not a book of the Bible but is a collection of writings believed to be from several Egyptian scrolls discovered in the early nineteenth century during an archeological expedition by Antonio Lebolo. Members of the Church of Jesus Christ of Latter-day Saints (LDS Church) purchased the scrolls from a traveling mummy exhibition on July 3, 1835. They were then translated into English by Joseph Smith. According to Smith, the book was "a translation of some ancient records, claiming to be the writings of Abraham, called the book of Abraham, written by his own hand, upon papyrus." The scrolls described Abraham's early life, his travels to Canaan and Egypt, and his vision of the universe and its creation. The book of Abraham is relevant to the subject of *Angel Communication Code* for its description of what is known as Abrahamic astronomy.[15] The book of Abraham chapter 3 describes an astronomical construct that is not easy to understand.

The book of Abraham uses "the earth upon which thou standest" (Abraham 3:3, 5–7) as its point of reference. It mentions various heavenly bodies, such as "the stars" (Abraham 3:2), among which is Kolob (Abraham 3:3–4). These provide a fixed backdrop for the heavens. Among the stars are various bodies that move in relation to the fixed backdrop, each of which is called a "planet" (Abraham 3:5, 8) or a "light" (Abraham 3:5–7). Each of these planets is associated with "its times and seasons in the revolutions thereof" (Abraham 3:4). These lights revolve around a fixed reference point, "the earth upon which thou standest."

It is believed that the astronomy revealed to Abraham was intended to take conceptions of the cosmos already familiar to the ancient Egyptians and replace them with a proper gospel understanding. Abraham was directed by God to teach this astronomy and also to teach the Word of God using astronomic references. This might explain why the book of Abraham contains a prescientific description of the cosmos rooted in the ancient world. This could only be successful if Abraham linked the cosmos to gospel truths in ways the Egyptians could understand.

The larger point here is that much was revealed to Abraham from messages directly delivered to him by three angels. Abraham has very strong links to our extraterrestrial communication code numbers. This is probably deliberate and not random or some fantastic coincidence.

Regarding King David

David was the second king of Israel according to the Hebrew Bible, and the number 2 is important to our search. Why does Matthew 1:1 choose to reference King David as an ancestor of Jesus? As cited previously, Jesus is not a blood descendant of King David. More has been written about David than about any other person in the Bible. The name David means "beloved." Sixty-six chapters within the books of the Bible are dedicated to David. This does not include the fifty-nine references to him in the New Testament. This guy was and is important to Christianity and the entire history of the nation of Israel.

David's father was Jesse. When it was time to choose the future king of Israel, David was not one of Jesse's seven begotten sons who stood before Samuel for selection. Maybe Jesse thought that since David was the youngest, there was no way he would be anointed king. Maybe it was because David was watching the sheep and Jesse did not feel it was necessary for him to be there. Nobody knows. For some reason, Jesse

[15] Wikipedia, s.v. "Book of Abraham."

overlooked the one whom God would anoint as the next king. If there is a lesson in this, it is that, often, the one we least expect is the one God decides to use.[16]

David left his home at a very young age to find his way in the world. He was a young child when Samuel (Israel's godly prophet) found him and took him in. At a very young age, David was made a shepherd and given the sole responsibility of tending the family herd. This was a very big responsibility, and it demonstrates that he was trustworthy and responsible at a very early age. It is said that he was in God's training ground to be made an important king one day. Then there is the story of the shepherd boy David killing Goliath the giant with a rock, a slingshot, and a lot of faith.

David was promised by God that his children would rule Israel forever. He is one of the most well-known figures in Jewish history. He is known by many titles: David the conqueror, David the pious man, David the sweet singer, David the shepherd, and David the penitent. It was during the time of King David when the word *Jew* came into existence.

Jews didn't start using the word *Jew* as a way to identify themselves until after 500 BCE. In the Hebrew Bible and the Torah (the text that is most sacred to Jews). Before that, the term used most often to describe God's people was "the sons and daughters of Israel," or "Israelites," or "Hebrews."

Ancient Israelites originated roughly in the territory of modern Israel, also known as the ancient Levant or ancient Canaan, sometime before 1000 BCE. These people were united by a sense of shared ancestry, myth, ritual, and history. Ancient Israelites believed that they were descendants of three people: Abraham, his son Isaac, and his grandson Jacob (later named "Israel" in the Bible). What bound ancient Israelites together was the Exodus narrative from the second book of the Hebrew Bible.

The golden age of ancient Israel refers to the time of King David and his son and successor King Solomon. David reigned for forty years, from 1011 BC to 971 BC. Solomon, David's son and successor to his throne, reigned from 932 BC to 893 BC. King David is associated with founding the city of Jerusalem. It was David who brought the Ark of the Covenant to Jerusalem in the seventh year of his reign. King Solomon is associated with the construction of the first great temple of God in 957 BCE. The dates vary depending on one's source of historical data.

Ancient Israelite society was divided between twelve tribes, ten of them in the north of the region and two of them in the south. In 722 BCE, the ten tribes of northern Israel were conquered by the ancient Assyrian Empire, and only the tribes of the south, in the tiny kingdom of Judea, remained as their own self-ruling political territories.

The time of King David and King Solomon was a very brief golden age, lasting only two generations. Most of the story of the ancient Israelites is a story about a tiny people among other superpowers trying to maintain their identity and their relationship with God while surrounded by other competing gods and more successful peoples. The situation remains similar to this day.

In the sixth century BCE, the Empire of Babylonia conquered the tiny kingdom of Judea and seized the Ark of the Covenant. This conquest could have ended Jewish history. When the Babylonians conquered ancient peoples, not only did they destroy buildings and plunder wealth, but also they exiled the people

---

[16] Clarence L. Haynes, "Who Was Jesse in the Bible?," Biblestudytools.com, July 29, 2022.

most responsible for creating and maintaining the local culture. In 586 and 587 BCE, (mostly elite) Jews were exiled to Babylonia. Some Jews also fled to Egypt. This time could be considered the beginning of the Jewish Diaspora, a term derived from the word *dispersion*, meaning "the spreading out of Jews across the Middle East."

To this very day, David is remembered by Jews everywhere, and they still chant his psalms in prayer in times of both joy and sorrow.[17]

David is often credited with writing the book of Psalms; however, he is not the only author of the psalms. In fact, of the one hundred fifty psalms, David is identified as the author of only seventy-five. He is specifically cited as the author of seventy-three in the titles of the psalms. In addition, he is credited as the author of two psalms by writers in the New Testament. Psalm 2 is attributed to David in Acts 4:2, and Psalm 95 in Hebrews 4:7. David's psalms express a heart devoted to God, yet he is known to be a frequent sinner and was punished by God by the death of his perfectly healthy, but illegitimate, first child. The child's death was prophesied ahead of time as punishment for David's sins.[18]

Psalm 23, the Psalm of David, is arguably one of the most-known writings within the Holy Scriptures:

> The Lord is my shepherd; I shall not want. He maketh me to lie down in green pastures:
>
> He leadeth me beside the still waters. He restoreth my soul: He leadeth me in the paths of righteousness for His name's sake. Yea, though I walk through the valley of the shadow of death, I will fear no evil: for Thou art with me; Thy rod and Thy staff, they comfort me. Thou preparest a table before me in the presence of mine enemies: Thou anointest my head with oil; my cup runneth over. Surely, goodness and mercy shall follow me all the days of my life: and I will dwell in the house of the Lord forever.

This psalm has been referenced, cited, or otherwise quoted in almost every conceivable situation where people are in some sort of life-threatening peril and need hope. Many soldiers are known to have carried Psalm 23 into battle or placed it next to their hearts before going into a seemingly hopeless and life-threatening situation. It is often recited at funeral services and during last rite ceremonies in hospitals and nursing homes. There was actually a time in recent history when every schoolchild would learn it and say it as part of a daily routine. It is sad that those days are essentially gone because of pure politics. The politics, however, do not negate the spiritual impact made by prayers and Bible messages. In *Angel Communication Code*, we are seeking a code within the messages that will allow us to communicate with the community of the universe while still keeping faith in God and the bigger message. This is not a simple task.

Psalm 23 is appreciated and recited by nearly every person who claims to be Christian and also by some who make no such claim. Ironically, Psalm 23 has been recited by political figures who were instrumental in keeping God's Word out of our classrooms and the Ten Commandments out of our courtrooms. The hypocrisy of politics is as grotesque as it is ancient.

David was not always obedient to God's Word, and there were times when he sinned greatly. He always recognized his sins and prayed to God for forgiveness. David was said to be a man with a regenerated heart.

---

[17] Dovie Schochet, "The Story of King David in the Bible," Chabad, accessed November 24, 2023.

[18] Jeffrey Kranz, "Who Wrote the Psalms?," Bible Overview, October 12, 2018.

Let us look at the Star of David.

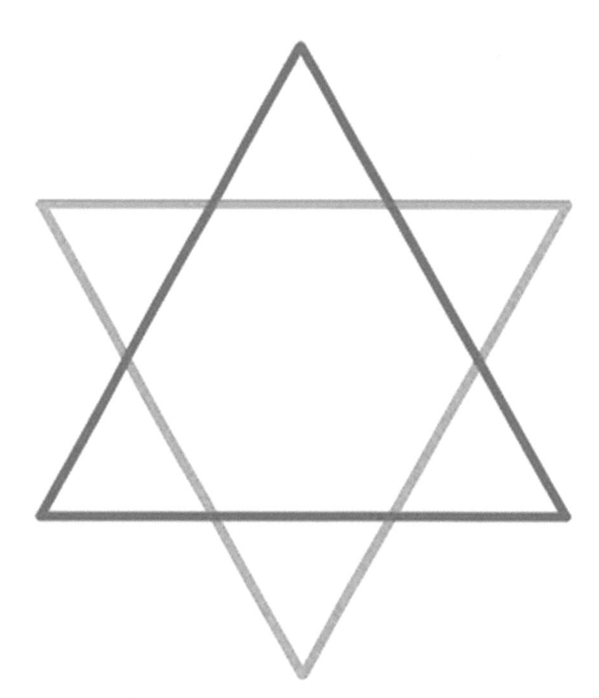

The Star of David is a six-pointed star, which represents peace and harmony in Buddhism. Ancient alchemists believed it symbolized nature. It is two equilateral triangles that are extremely significant to this ET communication code project.

The Star of David, before it was called the Star of David, appeared thousands of years ago in the cultures of the East, where it is still in use to this day. What Christians identify as the Star of David was a common symbol in pagan traditions. It was also in first-century churches and is also found Muslim culture.

In a Hebrew context, it is called the "Shield of David." The term was first mentioned in the Babylonian Talmud, not just as a symbol, but also as an alias for God (Pesachim 117b). The shield idea is derived from Jewish folklore. It is said that the star emblem decorated the shields of King David's army. In addition, Rabbi Akiva chose the Star of David as the symbol of Bar-Kochba's ("son of the star") revolt against the Roman emperor Hadrian.

The Star of David morphed into a distinctly Jewish symbol in the mid-fourteenth century, when the Roman emperor Charles IV granted the Jews of Prague the right to carry a flag. They chose the Star of David as the symbol for their flag. From Prague, the use of the Star of David as an official Jewish symbol spread. This was the start of the movement to find Jewish sources that traced the symbol to the House of David.

The "Star of David" hexagram has been used in India for thousands of years. It is commonly found on ancient temples and in daily use today. In Buddhism it is used as a meditation aid to achieve a sense of peace and harmony. In Hinduism it is a symbol of the goddess Lakshmi, the goddess of fortune and material abundance.

Hexagrams are also prevalent if the world of alchemy, which is the theory and study of materials and from where the modern science of chemistry was born. Magical symbols were commonplace in ancient theory, and alchemists incorporated the six-pointed star into their language of signs and symbols. An upright triangle symbolized water, and an inverted triangle symbolized fire. Used together, the two symbolized harmony between the opposing earth elements. In alchemical literature, the hexagram also represents the "four elements" and the concept that all matter in the world, everything that exists, is made up of them: air, water, earth, and fire. Alchemy took the concept from the classical Greek notion that masculinity is wisdom, and femininity is nature. Man is philosophy and woman is the physical world.

The illustration below, entitled *The Philosopher Examining Nature*, is take from a 1749 alchemical text from the Sidney Edelstein Collection at the National Library of Israel. It depicts a man holding a lantern (wisdom) while following a woman holding a hexagram (nature). They are the two keys that are said to unlock the secrets of existence.

The following is taken from the National Library of Israel's website:

> The Edelstein Collection covers all areas of the history, philosophy, and sociology of science. It includes sources as well as secondary literature, periodicals, and a reference section. Most books are in English, but there are also works in French, German, Latin, Spanish, Portuguese, Italian, and Hebrew. The books in this collection deal with the whole spectrum of thought between the exact and the occult, from magic to quantum mechanics, from chiromancy to cognitive sciences, from astrology to space technology.

The Star of David symbol is referred to as the Seal of Solomon in the Islamic world. It is incorporated into numerous mosques around the world. It was also found on the Moroccan flag until 1945, at which time the Moroccan flag was modified to a five-pointed star (pentagram)—when the six-pointed star became the emblem of the Zionist movement. The use of the Star of David has diminished throughout the Islamic world as well for the same reason.[19]

---

[19] Sharon Cohen, "The Story of the Star of David," National Library, May 19, 2021.

All this information about Abraham and David is a good enough reason for the first verse of the first chapter of the first book of the New Testament ([1] Matthew 1:1 = 3) to direct our attention to Abraham and David through the genealogy of Jesus via three distinct fourteen-year time segments. Now let us dig even deeper and consider the two-dimensional Star of David's incorporation into the three-dimensional Metatron's Cube as it applies to the subject of *Angel Communication Code*.

The theme of *Angel Communication Code* is about recognizing, understanding, and using the clues that may have been provided to us by angels or ETs, or both, that point us to a connection with and a way to communicate with other intelligent life in the universe. The trail now leads us to something called the Metatron Cube, which incorporates the Star of David.

Metatron is said to be a very special and powerful archangel. Archangels are called "the great heralds of good news" because God sends them to deliver the most important of His messages to humankind. These are the angels who directly communicate and interact with us when instructed to do so by God. Archangel Metatron is said to be the archangel of life who oversees the flow of energy in a mystical cube known as Metatron's Cube. Clues from angels about how to communicate with life-forms in other places of the universe are what we seek in *Angel Communication Code*. Now we are looking at a giant clue we were led to by looking closely at the importance of Jesus's genealogical reference to King David in Revelation, and the Star of David.

Metatron's Cube is said to contain every shape that exists in the universe that God has created. These shapes are the building blocks of all physical matter. They are called Platonic solids, of which there are only five:

1. tetrahedron—the four triangle faces that are associated with the element of earth
2. cube—the six square faces that are associated with the element of water
3. octahedron—the eight triangle faces that are associated with the element of air
4. dodecahedron—the twelve pentagram faces that are associated with the element of fire
5. icosahedron—the twenty triangle faces that are associated with the spirit.

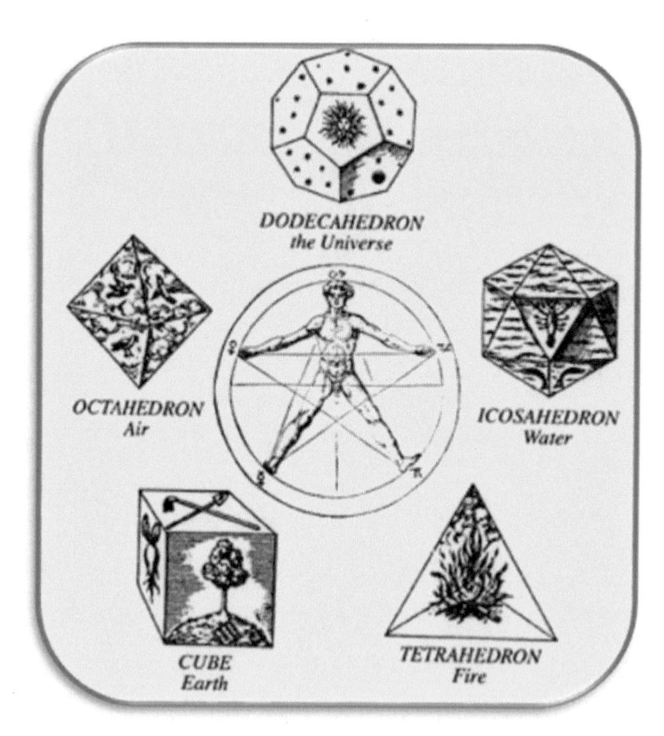

These three-dimensional shapes appear throughout all creation, from crystals to human DNA.

The Metatron's Cube was first identified by the Italian mathematician Fibonacci, and its three-dimensional graphics are considered to have roots in what is known as nature's first pattern. Fibonacci also discovered the Fibonacci sequence, and the golden ratio, previously discussed in *Angel Communication Code*.

There are important clues other than the numbers 2, 3, 5, and 7 that could be leading us to the key to the code. Look at the two primary shapes in geometry: the equilateral triangle and the circle. These two fundamental shapes are conspicuously used everywhere throughout history and all over the world as important and meaningful symbols. They are also the most prevalent shapes that UFO observers report.

The circle is the only defined geometric shape that is a single line that begins where it ends and has one radius and one diameter. It is also the only shape in geometry that has a component that seemingly never ends. That component is pi: the length of the circumference of the circle divided by the length of its diameter. This has been known for thousands of years.

The equilateral triangle has some unique properties compared with the other forms of triangles in that:

1. All three sides and all three angles are equal.
2. It is the strongest geometric shape for structural engineering.
3. The angle bisectors, the medians, and the perpendicular bisectors of the three sides coincide at a point. That point is the centroid of the triangle.
4. The incenter and the circumcenter coincide at the same point within the triangle. That point is the centroid of the triangle.

The Eye of Providence, or "All-Seeing Eye," is an ancient circle and triangle symbol used by numerous cultures throughout history. The Eye of Providence is a favorite topic among conspiracy theorists. It is everywhere but usually is very subtle and in the background. It shows up in countless churches and Masonic buildings all over the world. It is part of the Great Seal of the United States. The circle and triangle mean something to humanity with respect to ET links and must somehow be involved in our search for an ET message. The image below is the All-Seeing Eye on the US one-dollar bill.

Annuit coeptis means "providence favors our undertakings," and *novus ordo seclorum* means "new order of the ages." MDCCLXXVI = 1776, the year the Declaration of Independence was ratified on July 4. The All-Seeing Eye, the circle and triangle together, dates as far back as the ancient Egyptians and the Eye of Horus.

The Eye of Horus is one of the best-known symbols of ancient Egypt. Known also as the Wadjet, this powerful symbol is believed to provide protection, health, and rejuvenation. Because of its powerful protective powers, the Eye of Horus was popularly used by the ancient Egyptians, both the living and the dead, as amulets. Even today, the Eye of Horus continues to be used as a symbol of protection.

The Eye of Horus is not just a magical symbol but is also an example of the mathematical knowledge acquired by the ancient Egyptians. The Eye of Horus is a story unto itself.

Although ancient Egyptian civilization ended, the belief in the potency of the Eye of Horus continued, and this symbol is still used by many today. As an example, in Mediterranean countries, fishermen often paint this symbol on their vessels for protection. Additionally, many people still wear the Eye of Horus as jewelry, to protect themselves from the ill will of others. Moreover, the Eye of Horus is popular among occultists, and also conspiracy theorists, who view the Eye not only as a protective symbol, but also as a symbol of power, knowledge, and illusion.[20]

So, from where does this "eye in the sky" concept ultimately derive? The Wide Field and Planetary Camera 2 of the Hubble telescope recently returned some pictures of the Engraved Hourglass Nebula known as MyCn-18.

---

20 "Eye of Horus: The True Meaning of an Ancient, Powerful Symbol," Ancient Origins, November 18, 2018.

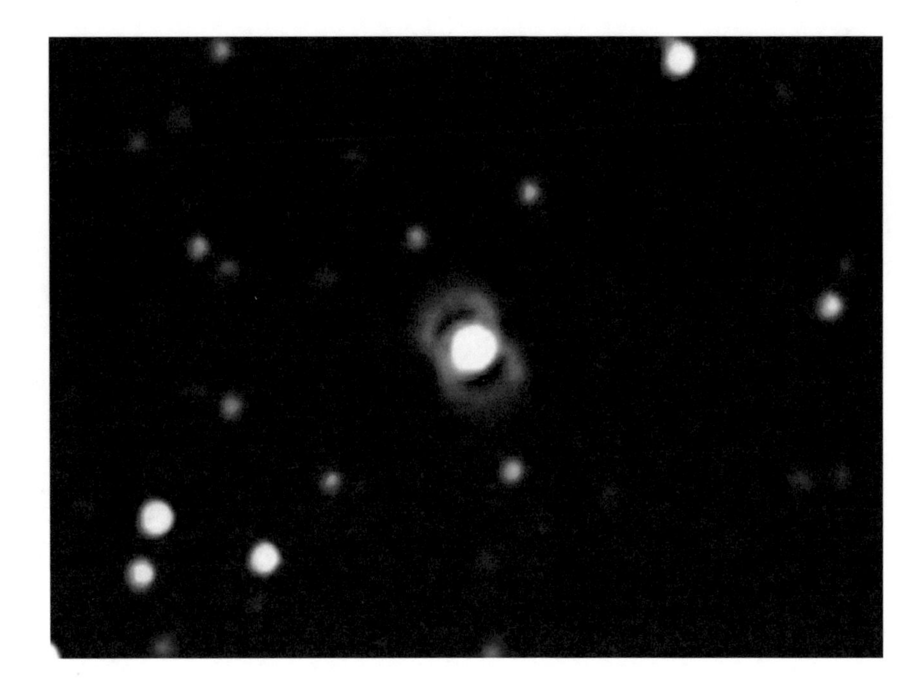

The closer the camera zooms in, the more the eye shows up.

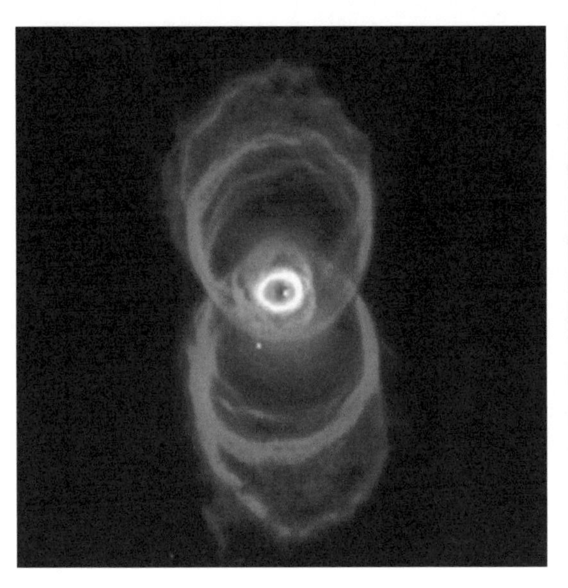

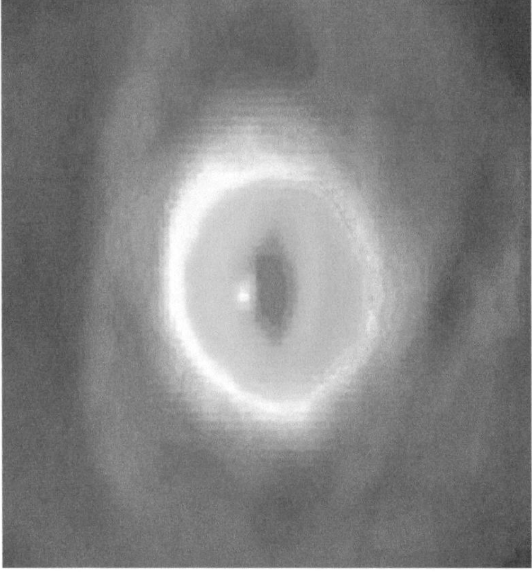

MyCn-18 is a young planetary nebula in the southern constellation Musca. It was discovered by Annie Jump Cannon and Margaret W. Mayall during their work on an extended *Henry Draper Catalogue*, which was compiled between 1918 and 1924. At the time, MyCn-18 was designated simply as a small faint planetary nebula. Much-improved telescopes and imaging techniques allowed the hourglass shape of the nebula to be discovered by Romano Coradi and Hugo Schwarz in images taken during 1991–92 at the European Southern Observatory.[21]

Eta Muscae is a multiple star system in the southern constellation of Musca. The system is located approximately 406 light-years away from the sun. The Greek work *Eta* is of huge significance in *Angel Communication Code* as you will soon see. There is something out there that looks very much like an eye in the sky, but how would the ancients have known this? Perhaps it is just a giant coincidence, or perhaps

[21] Wikipedia, s.v. "Engraved Hourglass Nebula."

it is not—and somehow the ancients were, with some help from extraterrestrials here on earth, able to observe this cosmic eye in the night sky. Thus the legend remains in service to this day.

Getting back to Metatron's Cube, following is a very common, readily available two-dimensional representation of it. Notice the patterns of circles and triangles.

Metatron is arguably the greatest of angels found in Jewish myths and legends. Metatron is not mentioned in the Hebrew Bible; however, his name appears briefly in several passages of the Talmud. His legends are predominantly found in mystical Kabbalistic (Jewish mystical) texts. He has been intermittently identified as the Prince (or Angel) of the Presence, as Michael the archangel, or as Enoch after his bodily ascent into heaven. He is commonly described as a celestial scribe recording the sins and merits of human beings, as a guardian of heavenly secrets, as God's mediator with humankind, as the "lesser Yahweh," as the archetype of humankind, and as one "whose name is like that of his master."[22]

Metatron is mentioned in passages of the Babylonian Talmud, in mystical Kabbalistic texts, and in the apocryphal books of Enoch. The Talmud is a central text of Judaism consisting of discussions and commentary on Jewish history, law, and customs. For a long time, the Talmud was passed down as oral tradition, until it was compiled and recorded in the second century AD as a document called the Mishnah. Commentaries on the Mishnah were then written down, making up the Gemara, the second part of the Talmud. The Babylonian Talmud was completed in the fifth century AD.

The Talmud is not considered a sacred work in Christianity, and though some teachings from it might be compatible with Christian teachings, the entirety of the New Testament had been completely written and largely compiled by the time the Mishnah was written down and made canon, before the Babylonian Talmud came into existence.

It is said that Metatron was once a human named Enoch, who is mentioned in the book of Genesis as a man who walked faithfully with God. Enoch was taken to heaven without dying, where he was transformed into the angel Metatron and placed on a throne next to God's throne. He became second only to God in terms of power, wisdom, and glory, and all the other angels answered to him.

Metatron is said to be part of a select group of angels who are allowed to look upon God's face. He is a heavenly scribe, as well as an advocate or heavenly priest for the people, and a mediator between Israel and God. One legend says it was Metatron and not Moses who led the Israelites through the wilderness.

Metatron is never mentioned in the Bible. However, Enoch is included in two passages. Genesis 5:18–24 tells the (brief) tale of Enoch's life:

> When Jared had lived 162 years, he became the father of Enoch. After he became the father of Enoch, Jared lived 800 years and had other sons and daughters. Altogether, Jared lived a total of 962 years, and then he died.
>
> When Enoch had lived 65 years, he became the father of Methuselah. After he became the father of Methuselah, Enoch walked faithfully with God 300 years and had other sons and daughters. Altogether, Enoch lived a total of 365 years. Enoch walked faithfully with God; then he was no more, because God took him away.

Hebrews 11:5 yields even more information:

> By faith Enoch was taken from this life, so that he did not experience death: "He could not be found, because God had taken him away." For before he was taken, he was commended as one who pleased God.

So, Metatron the angel is not in the Bible, but could he still exist? Not if he was first the human named Enoch. Humans and angels are entirely different beings, and humans do not become angels. But what if he wasn't a human first at all? Metatron and his cube are the work of an anonymous "mystic" in the same

---

[22] *Encyclopedia Britannica*, s.v. "Kabbala—Jewish Mysticism."

way the hierarchy of angels accepted by Christianity was the work of the unknown mystic they called Pseudo-Dionysius.[23] Those facts are very important as it means these things were constructed for a purpose by unknown entities, and that purpose may very well have been to create a coded message about ET communication.

Looking at sacred geometry leads a person closer to an understanding of how God has structured the physical world around us. Patterns emerge that point to unity and connection to the divine forces that created them. Ancient geometric codes underlie seemingly disparate things, showing the parallels between patterns in snowflakes, seashells, flowers petals, the corneas of our eyes, the DNA molecule that is the building block of life, and the galaxy itself in which Earth resides.[24]

In his book *Beautiful Schools*, Ralph Shepherd sees Metatron's Cube as a symbol of how God made shapes fit together throughout creation and how He designed people's bodies and souls to fit together. "The cube represents the three-dimensionality of space. Within the cube lies the sphere. The cube represents the body with our third-dimensional reality of manifested thought. The sphere within represents the consciousness of spirit within us, or as is commonly known, our soul."

The cube is said to be an image of God's energy flowing through Metatron to all the many parts of creation, and Metatron works to ensure that this energy flows in the proper balance so that all aspects of nature will be in harmony. This is consistent with Adrian Bejan's constructal law theory and the divine ratio previously discussed. It shows how all shapes and structures in nature arise to facilitate flow and balance.

We read in Rose Vanden Eynden's *Metatron: Invoking the Angel of God's Presence*, "Metatron's Cube helps us realize the harmony and balance of nature .... Since it depicts equilibrium in the six directions represented within it ..., Metatron's Cube can be used as a visual focal point to connect with the archangel, or it can be used as a concentration tool for meditations that promote peace and balance. Place an image of the cube anywhere you wish to be reminded of the archangel's loving, balancing presence."

Vanden Eynden further states, "Ancient scholars believed that by studying sacred geometry and meditating on its patterns, inner knowledge of the divine and our human spiritual progression can be gained," It is believed that people can derive inspiration from Metatron's Cube in sacred geometry and also use it for personal transformation.

Metatron is believed to use his cube for healing and clearing away lower energies. The cube is said to spin clockwise and to use centrifugal force to push away unwanted energy residue. The archangel Metatron has insights into the workability of the physical universe, which is believed to be composed of not only atoms but also thought energy. Metatron can help you work with universal energies for healing, understanding, teaching, and even bending time. Metatron's Cube is said to be a means of "transformation" to help humans to listen with their hearts and minds so that they can be connected to the infinite universe. Archangel Metatron's Cube contains many geometric symbols for the unity of the finite and the infinite.[25] The concepts of universal energies, connecting with the infinite, and bending time are extremely significant clues linked to the finding of a code that can lead us to establishing two-way communication with ETs.

---

23 Alyssa Roat, "Who Is Metatron?," Christianity.com, March 20, 2020.

24 Rose Vanden Eynden, *Metatron: Invoking the Angel of God's Presence* (Woodbury, MN: Llewellyn, 2008).

25 Whitney Hopler, "Archangel Metatron's Cube in Sacred Geometry," Learn Religions, August 31, 2021.

Let us now examine the critical mathematical links to our 2, 3, 5, and 7 ET code numbers and Metatron's Cube. A "Platonic" solid is a convex, regular polyhedron in three-dimensional Euclidean space. A "regular" polyhedron has faces that are identical in shape and size, and the same number of faces meets at each vertex. There are only five such polyhedrals that exist as far as we know.[26]

This line of thinking starts with the Star of David, which is *two* equilateral triangles. One triangle is pointing to the earth, and the other is pointing to heaven or the universe, as shown again below.

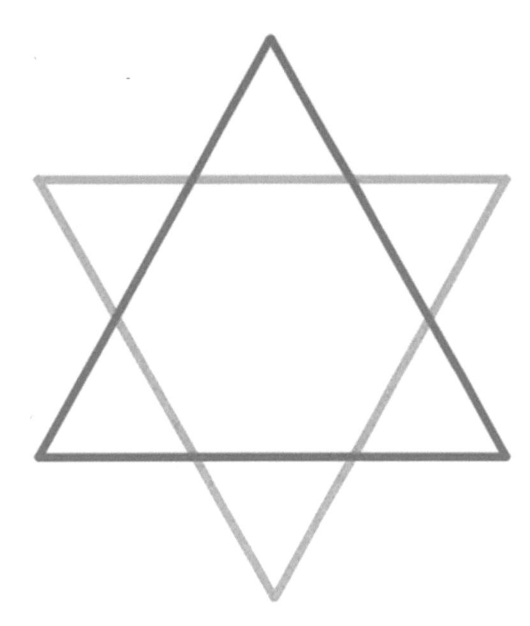

The two triangles are centered on each other, and given that this is the case, their independent concentric circles would be too. Their inscribed circles (two circles) become one shared circle. The same is true of their circumscribed circles.

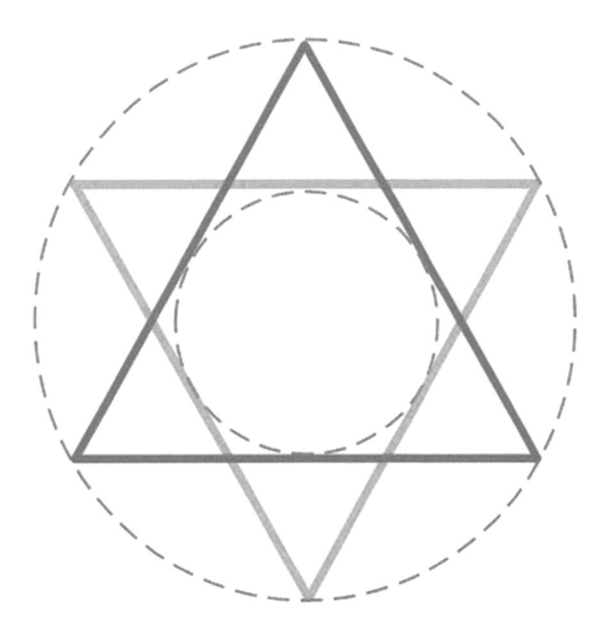

Below is the Star of David with its inscribed and circumscribed circles, laid over Metatron's Cube.

26 Wikipedia, s.v. "Platonic Solid."

Now let us examine this diagram with an eye toward our subject matter. Circles play a huge role in depictions of Archangel Metatron's Cube, but let us look at it in light of our key prime numbers and as a potential ET communication code key.

Metatron's Cube depicts seven equal circles that fit within the limits of the Star of David's two triangles (red circles).

Seven even circles of the same diameter fit within the boundaries of the Star of David. Notice there are three circle rows that lay out in three directions with the center circle shared three times. It is three rows of three circles with the number 2, the first prime number at its core, shared three times, as shown below.

When we move six of the circles to the vertices of the Star of David and keep the central "core" circle in place, it looks like this:

Two-thirds of each of the outer circles are now outside the Metatron's Cube, and the core remains in place. If we highlight the circles that reconnect the structure, it looks like this:

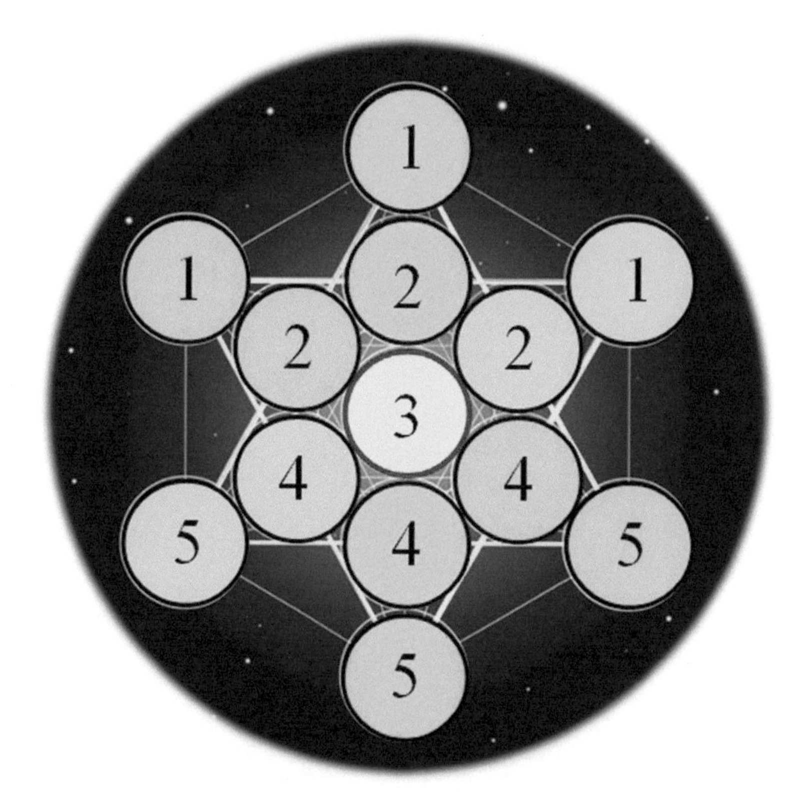

This is now three rows of five circles, which equals fifteen circles by using/sharing/counting the one center circle three times, one time for each five-circle row. It is three number 1 circles, three number 2 circles, three number 4 circles, and three number 5 circles, but only one number 3 circle, shared three times in the structure's core.

If we take this one step farther and add one circle that circumscribes the vertices of the two triangles and another circle that circumscribes the circle whose center points are the triangle vertices (two green circles), we get a total of seventeen (15 + 2) circles defined using the Star of David triangles as control points. This picture is shown below. The number 17 is a number of great significance in the Bible. It is also the seventh prime number.

This analysis identifies something very curious, namely that a band of space lies between the (green) circumscribed circle defined by the limits of the Star of David triangle's vertices and the (green) circumscribed circle at the outermost tangents to the Metatron's Cube circles, whose center is the Star of David's triangle vertices. What could be the significance of this region of space within Metatron's Cube?

Let us call it "green space" for purposes of discussion in *Angel Communication Code*.

There is more to this analysis on a tangible, earthly, and human level. Let us go back to basics and consider Leonardo Da Vinci's Vitruvian Man (shown below) as it relates to this geometric analysis of Archangel Metatron's Cube and the Star of David.

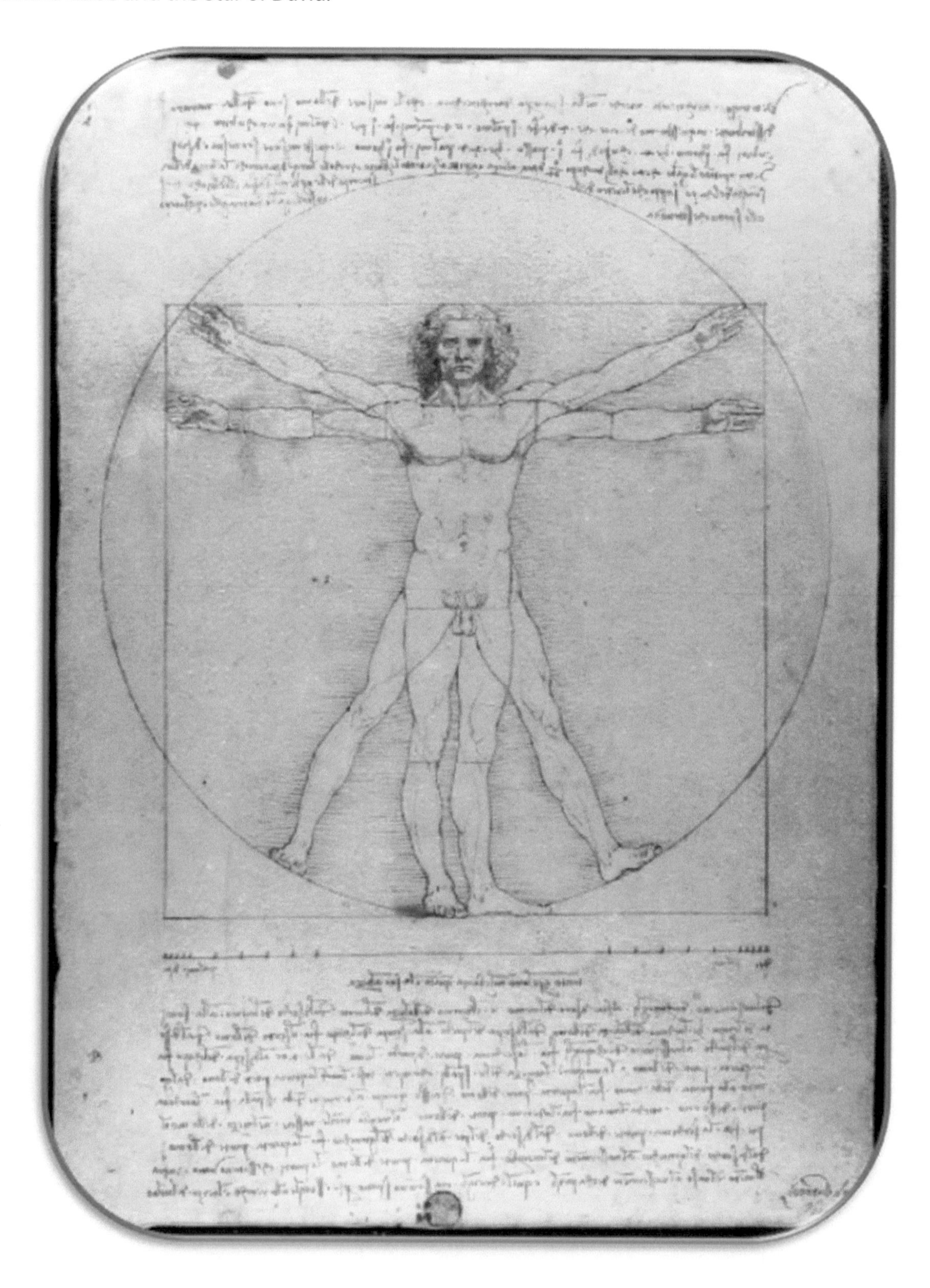

Da Vinci believed, as did many other people of philosophy and science in those days, that the human body contained the universe, as does Archangel Metatron's Cube; however, this link between the two has never been made until now. Da Vinci's Vitruvian Man is an effort to satisfy the symmetric requirements for harmony laid out by Vitruvius. Both these men subscribed to the idea that within the geometry of the human body lies hidden the mystery of the universe. The Vitruvian Man was, in a sense, the measure of all things.

Christians such as Augustine of Hippo (354–430) applied similar insights to make sense of the figure of the cross and the body of Jesus outstretched upon it. For Augustine, the cross was the key to charting the architecture of spiritual deliverance from sin, resulting in spiritual salvation. Considered as a microcosm or a "miniature world within a universe," the human body measures all things in its inherent firmness, functionality, and beauty. The human body is therefore a gateway to a universal reality.[27]

If we overlay the Star of David and "green space," it looks like this if we make the inner circle match up with the Vitruvian Man's reach:

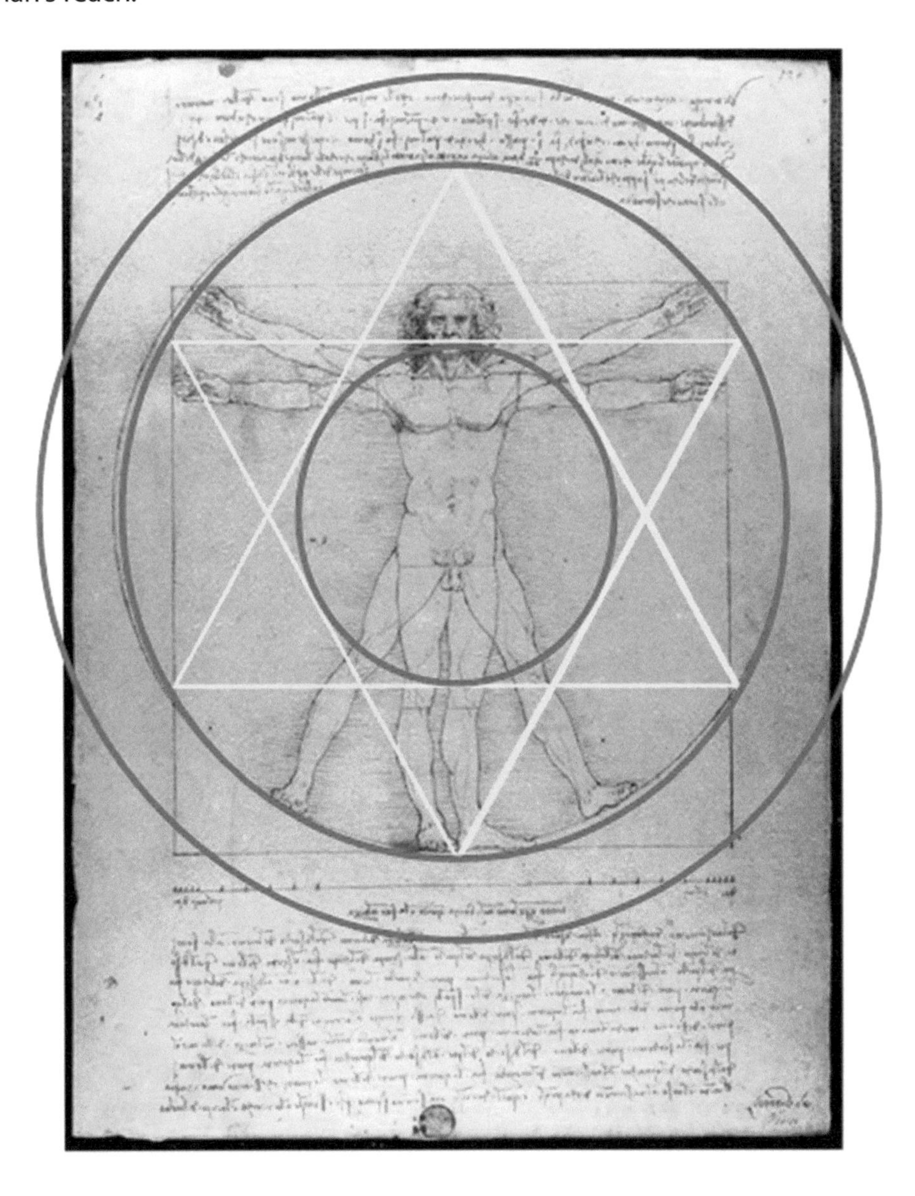

[27] Pablo Irizar, "Da Vinci's Vitruvian Man and the Measure of All Things," *Antigone Journal* (May 2021).

Notice that with the reach of the hands limited to the circle at the limits of the Star of David triangle vertices, the head is in the lower portion of the point of the small upper (yellow) triangle pointing to the heavens. It is cut off at the mouth by the baseline of the triangle pointed down toward earth. This implies no audio-linguistic communication to the cosmos. Only the brain / mind / "third eye" is in the small upper triangle created by the two yellow lines pointing up to the cosmos and the one yellow line of the other triangle pointing down to earth. Only the mind is able to contemplate the cosmos, but we cannot verbally communicate with the cosmos. If we scale the Vitruvian Man to make it such that the man touches the outer limit of the Metatron's Cube, the "green space," then it tells a different story.

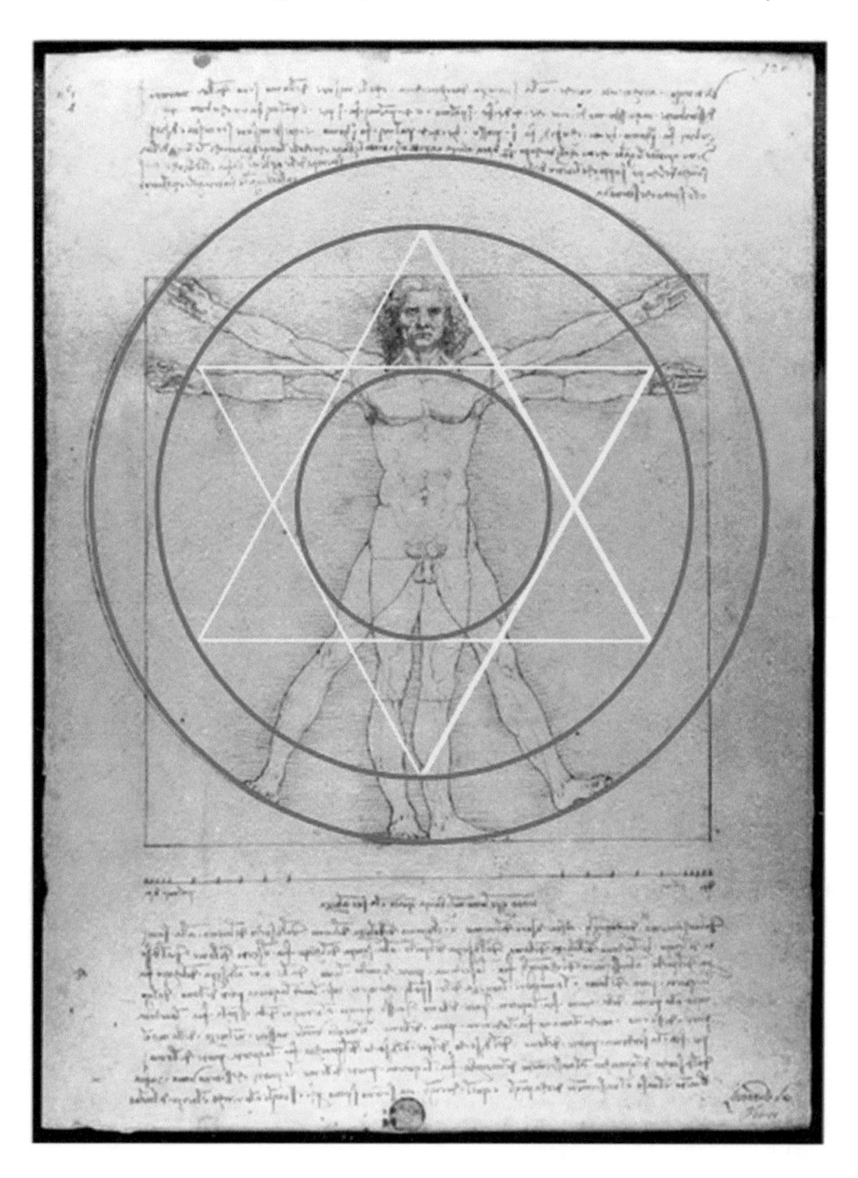

- The entire head and larynx (voice box) is as far up as it can go into the yellow triangle pointing to the cosmos.
- With the arms in the horizontal position, the hands can reach just beyond the limits of the Star of David into that "green space" we previously defined.
- With the arms held up toward the heavens to the same height as the top of the head, it defines the limits of our physical reach. Our physical bodies can go no further without leaving the mind behind.

What is this telling us? What is this green space? It tells us that all this stuff is somehow linked. It is pointing us toward the fact that the link to the cosmos lies in the use of circles, equilateral triangles, and the prime numbers 2, 3, 5, and 7. These are the keys to something we need to find and figure out.

There is no doubt that Leonardo Da Vinci was a man well ahead of his time. He buried many hidden messages into his works, and many believe there was an extraterrestrial influence at some point in his life, most likely during that one and only gap of time when nothing about his life was documented (much like the missing or lost years of the life of Jesus). Between the years 1476 and 1478, Da Vinci disappeared. His life did not otherwise have gaps in terms of documented coverage. Just before this time, he documented in his personal journal a story of being drawn to a cave while hiking. He was pulled in by the desire to experience the "wonder inside." Then he vanished for three years. And when he resurfaced, he could not explain his whereabouts and he started creating many of his advanced works with coded messages buried within.

We can take this even farther by looking at the critical circles in the combined images of Metatron's Cube, the Star of David, and the Vitruvian Man. We can link these to dimensions of consciousness as many currently believe they exist. And perhaps that is what Da Vinci already figured out, or was at least exploring and trying to convey. This will be discussed in much detail in the next chapter of *Angel Communication Code*, "Dimensions of Angels."

For now, let us circle back to our discussion, which began with an examination of a reason why the book of Matthew ([1] Matthew 1:1) may have cited David as the reference point for the genealogy of Jesus Christ back to Abraham. The sequence is from Abraham to David, then from David to the Babylonian exile, then from the exile to the birth of Jesus. The indicators (clues) that came out of that review caused us to the numbers 2, 3, 5, and 7. Now let us look more closely at the Babylonian exile part of this subject.

The Babylon Exile

The exile came after the reign of King Solomon, the son of David. Solomon became the third king of Israel by succeeding his father, David. David ruled for forty years, and so did Solomon. Solomon's forty-year rule was a time of peace and prosperity. After his death, the region was divided and not at peace.

The Babylonian exile describes a period in the history of ancient Israel well after the reign of Solomon. Modern interpretations of this period often are not as things really occurred. The Babylonian exile of the Jews has become proverbial. Protestant reformer of the sixteenth century Martin Luther speaks of a "Babylonian captivity of the Church." He does not refer to an actual exile but to of a period of spiritual enslavement of the church to the pontiff in Rome, a period that was soon to end.

Before the exile, Judah was a monarchy that had taken on the traditions of Israel, the tribal community once united under King David. It absorbed many of the pan-Israelite traditions, but it still was a commonwealth, a political entity with no other purpose than to exist, survive, and thrive as a political entity.

After the exile, Judah was politically rebuilt as a Persian satrapy, a semiautonomous administrative province, ruled by priestly elite who emigrated from Babylonia and whose views and attitudes were shaped by the religious blueprints for reconstruction drafted in the exile. These priests were at odds with the local population. They rigorously enforced separation from the mixed multitude of inhabitants of Judah and ruled on the basis of the Torah. This code of law was promoted by Ezra in the early fourth century BCE and served as the legal ideal of a theocratic state, that is, a state ruled by priests rather than kings. According to the later rabbis, the institution of the Torah as the basic law (in addition to which there must have been oral law traditions of various kinds) brought the earlier institution of prophecy to an end.[28]

The exile started with a two-stage deportation (597 BCE and 587 BCE) and ended with the conquest of Babylon by the Persian king Cyrus the Great in 538 BCE. The Babylonians, originating in what is now southern Iraq, rose to a position of power by the end of the seventh century by putting an end to the Neo-Assyrian Empire. Their king, Nebuchadnezzar II, had extended the empire to the east and to the west. On his way to control the trade routes to Egypt, he was confronted with the resistance of the kingdom of Judah. In 597 he conquered Jerusalem, exiled parts of the population including King Jehoiachin, and installed Zedekiah as puppet king. This event is reported in the Hebrew Bible (2 Kings 24:8–12) and also in the Babylonian Chronicle.[29]

Jews in Babylon also creatively remade themselves and refashioned their worldview during this time. In particular, they blamed the disaster of the exile on their own impurity. They had betrayed Yahweh and allowed the Mosaic laws and cultic practices to become corrupt. They considered the exile as proof of Yahweh's displeasure. During this period, Jewish leaders no longer spoke about a theology of judgment, but a theology of salvation. In texts such as Ezekiel and Isaiah, there were murmurings that the Israelites would be gathered together again, that their society and religion would be purified, and that a better and unified "Davidic" kingdom would be reestablished.

This was a time of resurgence in the Jewish faith as the exiles looked back to their Mosaic beginnings in an attempt to resuscitate their original religion. It is believed that the Torah was finalized around this period, or shortly afterward, and that this was also the time when it became the central text of the Jewish faith. This profound revival of religious tradition was actually assisted by another accident in history. When Cyrus the Persian conquered Mesopotamia, he allowed the Jews to return home, sending them home specifically to worship Yahweh, the Hebrew name for God.[30]

The exile is a story that mirrors the story of the Christian faith in that it is about losing faith then repenting in life, then dying, then being resurrected. The exile story is about falling from grace, rethinking everything, and going back to basics to get back on the right path to rediscover God and the heavens. That seems like a philosophy modern society might want to reflect upon more closely while also considering that history has a habit of repeating itself. Perhaps this is why Matthew included a period in time (vs. another person) as a reference point for the genealogy of Jesus. Perhaps the Babylonian exile inclusion in the ET code clue basket is a reminder that we should be preparing for an impending "extraterrestrial exile" or the prophesied battle of Armageddon and the beginning of the thousand-year reign over the earth by Jesus Christ. We might be closer than we know to any of or all that, based on the evidence being revealed all over the world today.

There is much in this chapter pointing us to the key prime numbers of and geometric clues to a message to communicate with the cosmos. As we move forward, we'll find that it all begins to come together into something quite specific.

---

[28] "Babylonian Exile and Beyond," Boston University, http://www.bu.edu/mzank/Jerusalem/cp/exret.htm.

[29] Bob Becking, "Babylonian Exile," Bible Odyssey, May 31, 2022.

[30] Richard Hook, "The Jewish Temples: The Babylonian Exile" and "The Hebrews: A Learning Module from Washington State University," Jewish Virtual Library, reprinted by permission.

CHAPTER 6

Dimensions of Angels

There are the mathematical definitions of geometric shapes, such as the circle and the triangle, whose significance to the subject of *Angel Communication Code* is being revealed. Then there are the dimensions of consciousness that have no equations or geometry. There are seemingly two ways to establish communication with ETs. One way is direct and physical, and the other is by reaching them via dimensions of consciousness. We can't have a conversation about extraterrestrial communication in the context of angels, or in other any context for that matter, without talking about things such as parallel universes and "dimensions" of consciousness. This subject is all over the place structurally yet is fairly consistent in concept. There seem to be two general groups of people who study these concepts and think they understand them: those with a scientific background and those with a more transcendental way of thinking going back to the time of ancient Greece and even earlier. It has been stated in *Angel Communication Code* that the key to achieving two-way communication with ETs is to think differently and go back to basics in our search for a coded message and clues to the solution to the problem. Let us overlay the image of Metatron's Cube on that of the Star of David as presented in the previous chapter of *Angel Communication Code* and align it with a uniquely new and more basic model of dimensions. It would look like this:

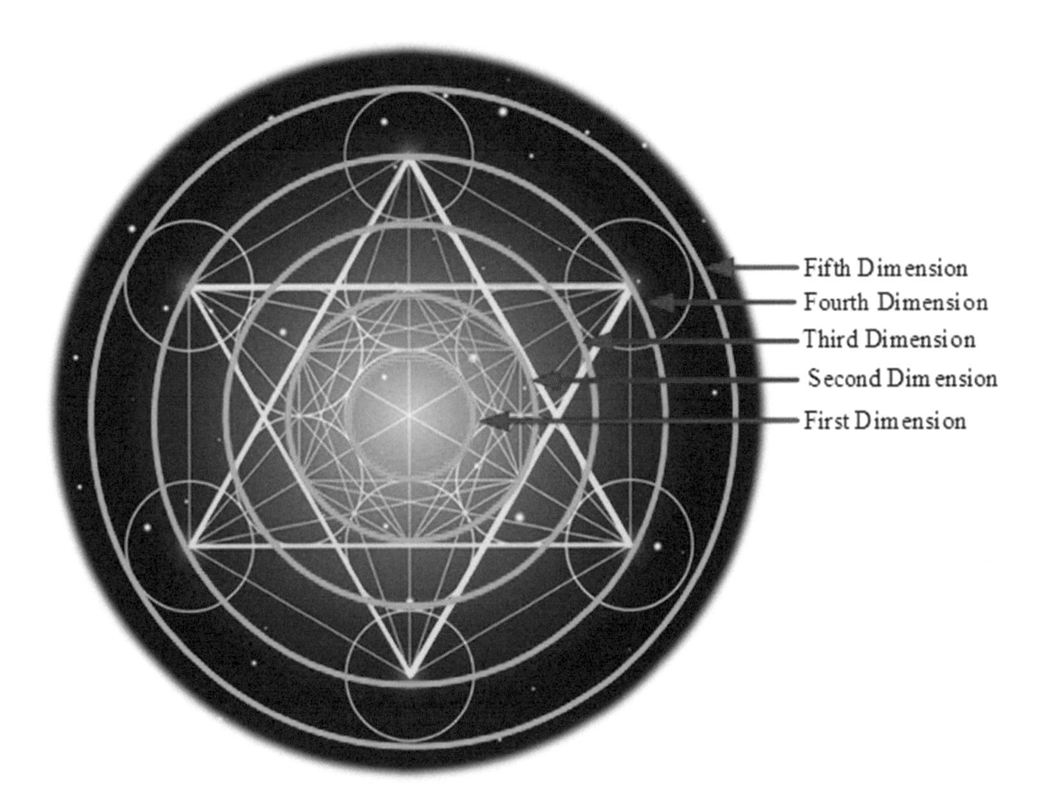

If you look carefully at the control points for each of the five green dimensional circles, from the most inner to the outer, you will notice the following:

1. the core of the Metatron's Cube diagram (the first dimension)
2. the inscribed circle of the Star of David (the second dimension)
3. the tangencies of the outermost set of circles that fit within the Star of David (it is also the tangencies of the outer Metatron's Cube circles centered on the vertices of the Star of David) (the third dimension)
4. the vertices of the Star of David triangles (the fourth dimension)
5. the outer tangencies of the Metatron's Cube circles centered on the Star of David's vertices (the fifth dimension).

The whole thing is beautifully symmetric. Now apply a third component, the Vitruvian Man, and you get this:

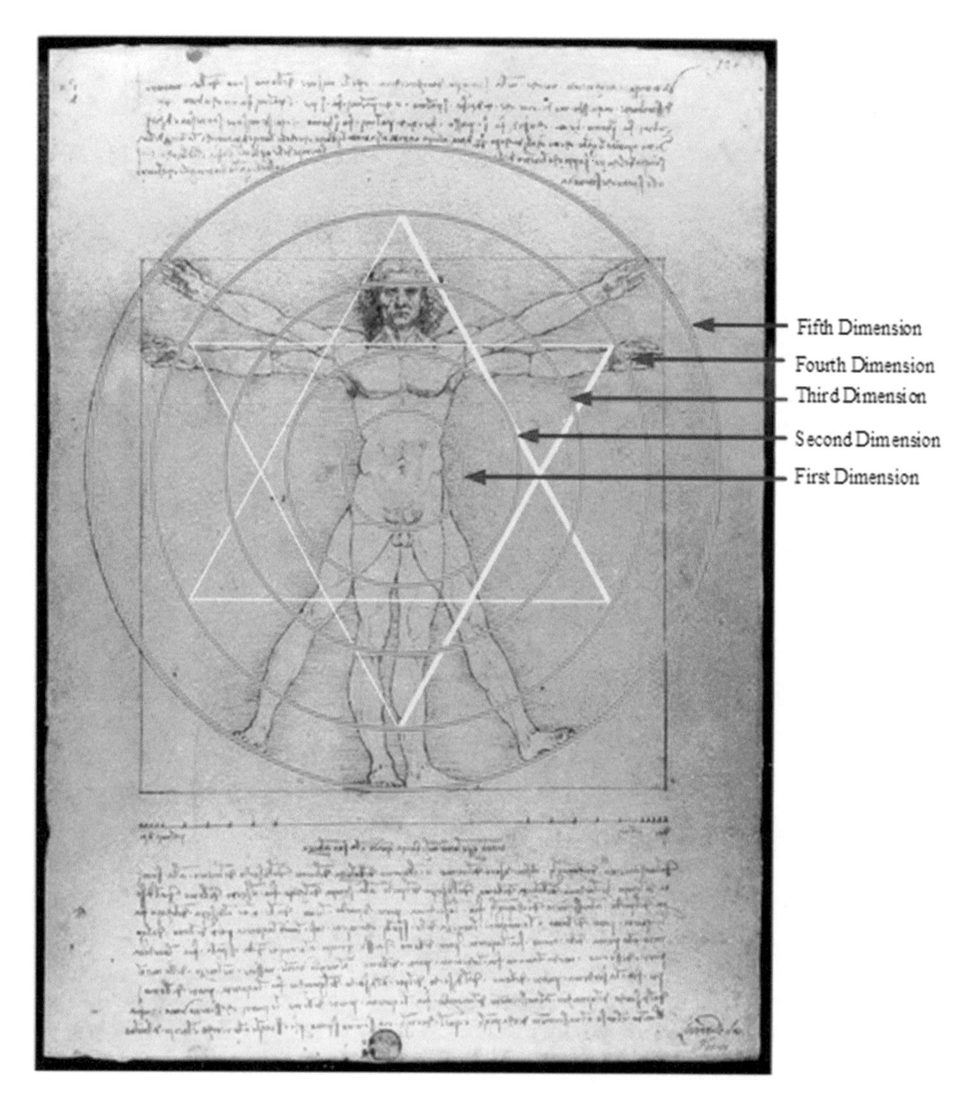

For the purposes of *Angel Communication Code*, we will call this the "Vitruvian dimension diagram" (VDD).

Compare all this to what we currently know about the dimensions of consciousness. It shows that humans can reach as far out as the outer edge of the fifth dimension in our physical form. After that lies the sixth dimension and so on. The journey into the universe then becomes one of pure consciousness, leaving our physical forms completely behind. This representation is consistent with, and seamlessly brings together, the following:

1. Archangel Metatron's Cube
2. the Star of David
3. Da Vinci's Vitruvian Man
4. the first five dimensions of consciousness
5. prime numbers 2, 3, 5, and 7.

There is a link, and that link has five components. This link, which is being presented and made for the first time in history here in *Angel Communication Code*, clearly has a purpose that leads to communication with the universe and all that dwell within it.

Since the time of Isaac Newton and even before, science has been driven by the belief that all things in the universe can be modeled using quantities that can be calculated using mathematical laws and the laws of physics. This concept is called the principle of reductionism in modern times. Reductionism in generic terms implies that the universal reality is understandable and therefore reachable. It implies that human beings, through the power of their minds and senses, may be able to understand the nature and origin of all phenomena in the universe. It is a grand concept that is not provable; however, it has provided an underlying strategy for scientific research. As scientists have gone from one discovery to the next, their faith in the universal applicability of this principle has grown stronger and stronger. The principle of reductionism does have its consequences, though. It reduces the universe to a mechanism operating according only to mathematical laws, thereby reducing individual humans to complex mechanisms whose thoughts, will, and feelings correspond to nothing more than identifiable and quantifiable chemical and electrical reactions between molecules.

Reductionist thinkers must recognize that they do not possess the exclusive right to knowledge of biological life, angels, ET existence, or our understanding of the universe. Alternative viewpoints deserve as much serious and scientific consideration as does the reductionist approach. Otherwise, scientist claims of unbiased and open-minded thinking become false claims, thus damaging the credibility of a so-called scientific approach to anything.[1]

It is a relatively simple effort to grasp the concept of the three dimensions of space: length, width, and height or depth. The concept of space is understood in a mathematically logical manner by defining a Cartesian system of measurements. Several of these "coordinate" systems have been devised, including liner, polar, and spherical coordinate systems. The most simplistic dimensional coordinate system is the Cartesian coordinate system introduced approximately four hundred years ago by French mathematician René Descartes.

Descartes's three dimensions of space, which humans perceive via our physical senses, can be extended conceptually to more than three dimensions. For each additional dimension, an additional dimensional point is needed. Geometry with more than three dimensions has been given the name "hyperspace." Hyperspace is a concept relating to higher dimensions and also to parallel universes and faster-than-the-speed-of-light interstellar travel. Its use in science fiction originated in the magazine *Amazing Stories Quarterly* in 1931, and within several decades it became one of the most popular imageries of science fiction, popularized by its use in the works of authors such as Isaac Asimov and E. C. Tubb and by media franchises such as *Star Wars*.[2]

---

[1] Madhudvisa Dasa, "Higher Dimensions in Science," *Origins*, January 28, 2014.

[2] Wikipedia, s.v. "Hyperspace."

Additional dimensions beyond the first three are said to be the measurable concepts of time and consciousness. It is believed that when a person's consciousness is expanded to include more than the planar dimensions described by Descartes, the person is not leaving this universe or somehow transported into another dimensional plane but, rather, is becoming aware of additional dimensions that already exist all around us. The individual was simply unaware of how to gain access to these dimensions. Just one additional dimension transforms the reality one experiences in ways that are impossible to describe before a person experiences it. The mind-expanding ways in which one's perceptions of reality are transformed become exponentially more complex with each additional dimension.[3]

Many believe that angels contact humans in the fourth dimension. They need not and cannot come any closer. This is the level of consciousness where dreams exist. When someone is said to have a vision-like experience, it is believed that this happens when the individual is in the fourth level of consciousness or the fourth dimension. For the most part, depending on one's reference source, the dimensions are generally described as follows:

The First Dimension

A reasonable geometric description of a one-dimensional object is a straight line, which exists only in terms of length and has no other geometric coordinates. What resonates most in this dimension is the seed of creation and life in its most basic form. It is the survival instinct. The first dimension is all about basic physical realities such as planets, air, water, and all physical elements. Science has studied this in great depth. Many believe that in that first instant following the (theoretical) big bang, the multidimensional nature of our universe actually became organized. Immediately following the big bang, there was a moment of singularity or a single point that contained all the energy in the universe. In that moment, the first dimension was created.[4]

Notice that the first dimension in the Vitruvian dimension diagram encompasses the human stomach (the basic need for food) and reproductive organs (our need to procreate).

The Second Dimension

The second dimension, geometrically, is the Cartesian *y*-axis (or height). It gives us things that become basic two-dimensional shapes such as circles and triangles. This is the realm of information. It begins with consciousness acquired by DNA information. Genetic codes resonate in the second dimension, as does the primal survival instinct. Science has also studied this extensively.

The Third Dimension

The third dimension is about the sort of intelligence that makes humans unique among all animals. Free will begins in the third dimension, which gives us the ability to make choices and decisions based on emotion and not survival. Philosophers and psychologists have studied the third dimension in great depth. The conscious mind of humans and the higher animal kingdom resonate at this level. The critical awareness of self as a separate entity begins in this dimension. In the third dimension, we experience duality. We experience the source energy as separate from ourselves so that we can observe that source objectively.

---

[3] Edward R. Close, "Dimensions of Space and Time," *Transcendental Physics* (October 25, 2021).

[4] "What Are the 1st, 2nd, 3rd, 4th, and 5th Dimensions?," Starseeds Compass, June 17, 2016.

Observing source energy allows for the manifestation of source energy into the physical realm. The double-slit experiment demonstrates how atoms behave as a wave of energy when they are not observed and as physical matter when they are observed. The awareness of the domain of three dimensions containing matter, energy, and conscious entities is only possible with the awareness of a moment of time existing in the fourth dimension, and the integrated experience of consciousness of the fifth dimension. Together, they comprise the experience of being a conscious, finite, self-aware being.

The Fourth Dimension

Many scientists believe that the fourth dimension is time, which governs the properties of all known matter at any given point. Along with knowing its position in the three lower dimensions, knowing an object's position in time is essential to plotting its position in the universe. The fourth dimension is the half step between the third and fifth dimensions. Thought forms that are close to manifesting into the physical are first experienced in the fourth dimension. The fourth dimension is the place where it is believed that angels make contact with humans. When we are in the fourth dimension, we can jump to higher dimensions more easily. The higher dimensions are where the deeper possibilities come into play. This is the dimension of the subconscious mind, where one can tap into and experience advanced psychic abilities. All mystics can tap into the fourth dimension. One also taps into it in some, but not all, dreams. The fourth dimension is where spiritual laws begin to manifest. Many spiritual teachers have been focusing on the fourth dimension over the past many years. There is a lot going on in the fourth dimension when we relate these things back to our Vitruvian Man diagram geometry (shown again below).

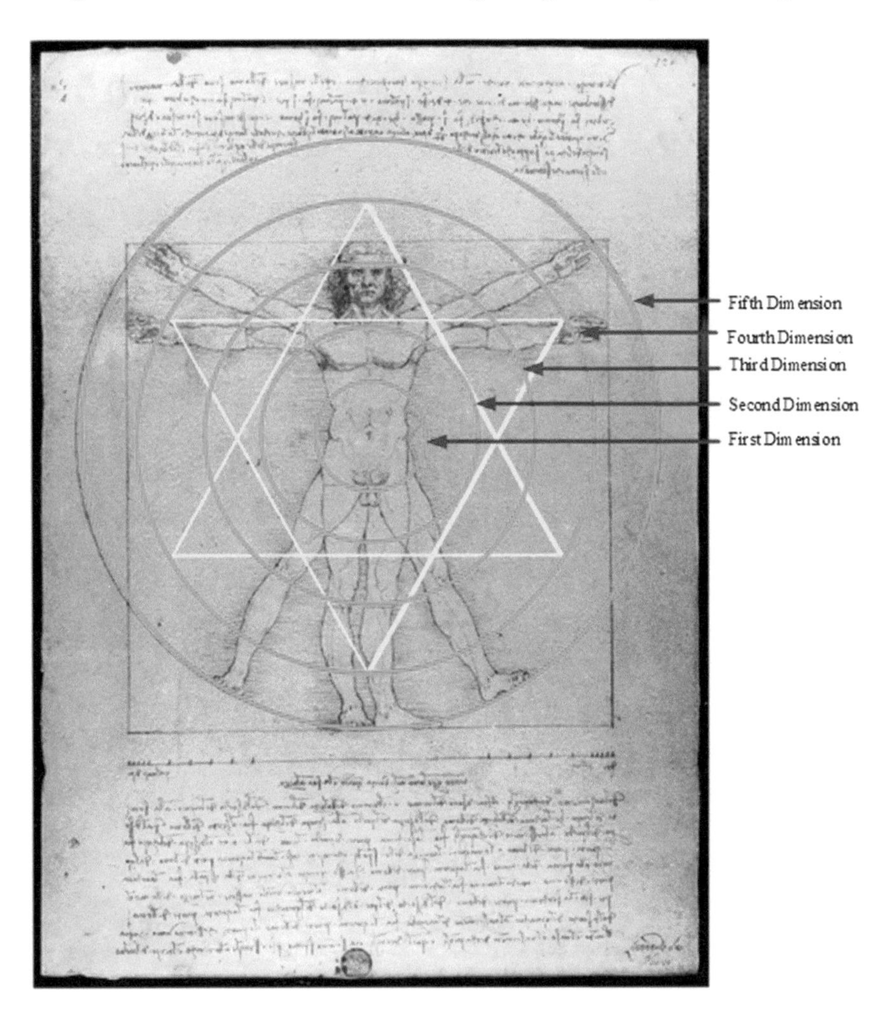

From the VDD with a focus on the fourth dimension circle, notice the following:

1. The fourth dimension ends at the limits of the Star of David triangles.
2. Regarding the head,
   a. sight, sound, smell, taste, and verbal communication all exist in the third dimension. Only the mind crosses into the fourth dimension.
   b. The line that ends the third dimension, where we cross into the fourth dimension, runs right through where the human pineal gland is located in the human brain. The pineal gland is also known as the third eye, which represents the third eye chakra, our transcendental connection to communication with the universe.
3. Regarding the feet, in any position, they are fully across the fourth dimension and bound at the limit of the fifth dimension. This is as far as our feet can get us through the dimensions without fully separating the mind from the body.
4. Regarding the arms,
   a. in the horizontal position, they run along the base of the Star of David triangle pointed down to earth. The arms reach just across the limit of the fourth dimension and into the fifth dimension but do not reach the limits of the fifth dimension. We are grounded.
   b. with the arms lifted entirely above the base of the Star of David triangle pointing downward to the same horizontal height as the head (mind), the body reaches fully across the fourth dimension and touches the limits of the fifth dimension. This is as far as we can reach through the dimensions without separating the mind from the body.

The Fifth Dimension

This is the limit of where the human body and mind can coexist. It is also the realm of spirit. This is where the first true witness, awareness, and understanding of spirit versus faith and belief in spirit is truly realized. The fifth dimension is the full understanding of God as a guiding light. Once you introduce a guiding light, you are no longer the center of your own creation, and duality dissolves into a higher purpose. Because duality starts to dissolve, there is no longer free will to do good or bad when you operate in the fifth dimension. You do what needs to be done. Angels are believed to operate in the fifth dimension, dropping to the fourth dimension only to deliver messages to humans. It is believed that approximately two-tenths of 1 percent of spiritual people have access to the fifth dimension. If we could see on through to the fifth dimension, we would see a world different from our own that would give us a means of measuring the similarity and differences between our world and other possible ones. The fifth dimension is where we experience source energy from a perspective of unity and where we have stopped asking ourselves, "What am I?"

In the fifth dimension, we perceive source energy for what it is. It is believed to be only consciousness and light. It is also believed to look like sine waves and fields of bright, dynamic, beautiful colors with dense inner parts that correspond with the third-dimensional body. The energy fields of all physical matter are visible in the fifth dimension. Because of this, there is rarely ever any empty space in the fifth dimension. It is possible to see physical bodies from the fifth dimension, but they are a bit obscured by the energy fields. In the fifth dimension, thought forms (waves) are visible, so manifestation is essentially instantaneous.[5]

We can take this idea even farther by incorporating sound and the "human pitch" of 110 hertz (Hz) or the 110 Hz phenomenon. The quantity 110 Hz is known to represent the human pitch. Buddhists and Hindus chant

---

[5] Ibid.

their mantras in this frequency in their chambers used for rituals and chanting. In these chambers, 110 Hz resonance enhances right-brain activity. The right brain is the center for art, poetry, sensuality, spirituality, feelings, imaginations, and innovations. The right mind is an intuitive gift, whereas the left mind is a faithful servant. Scientific studies have found that once able to access the right brain, a person becomes more problem-solving and less conflictual in nature. This was something known to our ancients, and so sacred to them that they built structures specifically to achieve this resonance.

Archaeoacoustics is a relatively new subdiscipline within archeology that studies the acoustics within archaeological sites and artifacts. Since many of the ancient cultures focused on the oral and therefore the aural, it is becoming increasingly recognized that the study of the sonic features within archaeological structures can enhance our understanding of these structures and the ancient cultures that created them. Archaeoacoustics is an interdisciplinary field that includes areas such as archaeology, ethnomusicology, acoustics, and digital acoustic modeling.

Emerging studies in archaeology by these new professionals, described by the Old Temples Study Foundation, suggest that sound and a desire to harness its effects may have been equally important to the visual aesthetics in the design of humankind's earliest ancient temples and monumental buildings. It has been found in such diverse places as Ireland, Malta, southern Turkey, and Peru. These structures may have been specially designed to conduct and manipulate sound to produce certain sensory effects, in particular, the generation of a 110 Hz sound, which may also be set up to be used as binaural beats.

A binaural beat is an illusion created by the brain when you listen to two tones with slightly different frequencies at the same time. Your brain interprets the two tones as a beat of its own. The two tones align with your brain waves to produce a beat with a different frequency. This frequency is the difference in hertz between the frequencies of the two tones. For example, if you were listening to a 440 Hz tone with your left ear and a 444 Hz tone with your right ear, then your brain would be hearing the 4 Hz tone delta. When you listen to binaural beats, your brain activity matches the frequency set by the frequency of the beat. This is called the frequency-following effect, which means you can use binaural beats to entrain your mind to reach a certain mental state.

The superior olivary complex located in the brain stem is the first part of the brain that processes sound input from both ears. The superior olivary complex synchronizes various activities of the many neurons in the brain. This complex responds when it hears two close frequencies and creates a binaural beat, which changes the brain waves. The synchronization of the neural activities across the brain is called entrainment. Not only is entrainment related to binaural beats, but also it is a common part of brain function. According to some researchers, when you listen to certain binaural beats, they can increase the strength of certain brain waves. This can increase or hold back different brain functions that control thinking and feeling.

Neurons in your brain use electrical signals to create thoughts, emotions, and behaviors. When neurons synchronize, this creates brain waves. Brain waves can be measured by a technique called electroencephalography (EEG).

As far as we know today, there are five different brain waves, as follows:

1. Delta, 1–4 Hz, which is the lowest frequency state and is linked to
    - deep sleep
    - healing and pain relief

- meditation
- antiaging—cortisol reduction / DHEA increase
- access to the unconscious mind.

2. Theta, 4–8 Hz, which has binaural beats benefits including
    - meditation
    - deep relaxation
    - creativity.
3. Alpha, 8–14 Hz, which causes your brain to be focused and productive. Alpha brain waves help you to
    - relax and focus
    - reduce stress
    - maintain positive thinking
    - increase your learning capabilities
    - easily engage in activities and the environment because you are in a state of flow.
4. Beta, 14–30 Hz, which is a higher-frequency brain wave and helps in
    - keeping your attention focused
    - thinking analytically and solving problems
    - stimulating energy and action
    - high-level cognition.
5. Gamma, 30–100 Hz, which helps with
    - increased cognitive enhancement
    - attention to detail, helping in memory recall
    - a different way of thinking, which is a sign of creativity.

Some studies have linked binaural beats to increased feelings of depression. Some people who listened to binaural beats experienced short bursts of anxiety, anger, and confusion.[6]

Having begun in 2008, a recent and ongoing study of the massive six-thousand-year-old stone structure complex known as the Hal Saflieni Hypogeum on the island of Malta is producing some fascinating results. The Hal Saflieni Hypogeum is a cultural property of exceptional prehistoric value and the only known example of a subterranean structure from the Bronze Age. The "labyrinth," as it is often identified, consists of a series of elliptical chambers and alveoli (where the lungs and the blood exchange oxygen and carbon dioxide during the process of breathing in and breathing out) of varying importance across three levels, to which access is gained by different corridors. The principal rooms distinguish themselves by their domed vaulting and by the elaborate structure of false bays inspired by the doorways and windows of contemporary terrestrial constructions. The structure is unique in that it is subterranean and was created by removing an estimated two thousand tons of stone carved with stone hammers and antler picks. Acoustically, low voices within the walls of this carved-out structure create eerie, reverberating echoes, and a sound made or words spoken in certain places can be clearly heard throughout all the structure's three levels. Scientists are suggesting that certain sound vibration frequencies created by the structure are at 110 Hz and that when sound is emitted within its walls, that sound alters human brain functions. Many of these sophisticated cultures created megalithic structures using a complicated aspect of archaeology known as corbelling: a system of oversailing row stones descending one by one, balanced by the weight of the stones being distributed equally. The mathematics involved in the design of this carved limestone

---

[6] "What Are Binaural Beats?," WebMD, April 12, 2021.

is present in many structures around the world. Whether it was deliberate or not, the people who spend time under its influence will resonate with the same frequency affecting their minds. Many structures all over the globe incorporate these phenomena.[7]

Physicist Michio Kaku says that the multiverse is always creating a kind of cosmic music. As this cosmic music resonates through eleven dimensions, it may be something like the "mind of God" described by Albert Einstein: "I want to know the thoughts of God. Everything else is just details."

On a smaller scale, everything has a unique sound, from mosquitoes to humans. A Stanford neurologist and a music composer have collaborated to create a tool to hear the human brain's signature sound. They call the instrument they use the brain stethoscope. By listening to the sound, we may soon be able to detect the brain's tone. One application of this tool is detecting when someone is having a silent seizure.

"The instrument, which is noninvasive and looks like a sweatband, straps onto a person's head and listens to the brain's electrical signals. With a push of a button, those signals are converted to sound that streams from a small speaker connected to the band. The thought is that doctors can 'hear' the tone of the brain—particularly if there is a seizure."

Another Stanford researcher created a tiny implantable medical chip the size of two grains of rice. Unlike other devices, this one is powered by ultrasound.

"A Swiss army knife of implantable devices, the chip can change its function to fulfill different biological needs. Its various modes are controlled by the same thing that fuels it. 'Ultrasound is both a power source and a way to communicate with the device,'" says Amin Arbabian.

With a tiny "harvester" module, the chip can convert ultrasound waves into electrical energy. Then the scientist can remotely communicate with the device using encoded commands. Thus, the implant can perform a whole host of functions as it monitors the body.

Reading about this incredible advance, one cannot help but think about the late Dr. Roger Leir. He performed surgical procedures on alleged alien abductees and claimed to find foreign objects whose descriptions sound much like the advanced implants described earlier.

Knowledge of sound waves and acoustics has long been part of human culture. Indeed, acoustic archaeologists have found some incredible things in structures dating back thousands of years. Some of the best-known examples are Stonehenge in the UK, Newgrange in Ireland, and the Hypogeum of Hal Saflieni on the island of Malta (cited previously). There, ancient people built incredible stone monuments that generate acoustic sounds, such as the Oracle Chamber in the Hypogeum. Inside this Oracle Chamber, sound resonates at precise frequencies around 110 Hz, as it does in Newgrange and other structures in Europe. These frequencies match the pitch that men's voices typically create when chanting. On account of the way the Oracle Chamber is spaced, speaking creates a standing wave pattern that echoes for seven seconds.

"It is said that standing in the Hypogeum is like being inside a giant bell. At certain pitches, one feels the sound vibrating in bone and tissue as much as hearing it in the ear," writes April Holloway for the website *Ancient Origins*.

---

[7] Robert Traynor, "The 110 Hz Phenomena," Hearing Health & Technology Matters, February 12, 2018.

Playing 110 Hz clearly has a specific effect on people's minds. A 2008 study by Dr. Ian Cook of UCLA found that for volunteers who listened to 110 Hz frequencies, the language center in their brains deactivated, switching to the right side.

"Findings indicated that at 110 Hz, the patterns of activity over the prefrontal cortex abruptly shifted, resulting in a relative deactivation of the language center and a temporary switching from left- to right-sided dominance related to emotional processing. People regularly exposed to resonant sound in the frequency of 110 or 111 Hz would have been 'turning on' an area of the brain that biobehavioral scientists believe relates to mood, empathy, and social behavior." Ancient peoples were doing this to facilitate an altered, or perhaps an elevated, state of consciousness.[8]

We are talking about going back to basics in our efforts to communicate with the community of the universe. The facts are that languages and dialects vary greatly; however, a circle is a circle and a triangle is a triangle everywhere in the universe. The laws of physics and mathematics work the same way everywhere in the universe, and everything in the universe vibrates and therefore has a frequency. Everything we are discussing is leading us down the trail that pulls all these things together with purpose, based on a dreamlike abstract vision leading to an initial focus and manipulation of the first three prime numbers, 2, 3, and 5, which lead to the fourth prime number, 7.

These alignments pull together and link physical dimensions, the circle, the triangle, Archangel Metatron's Cube, Da Vinci's Vitruvian Man, dimensions of consciousness, the human pitch, and the human physical equation. The alignments are unmistakable.

Never before has anybody put together or made these seemingly obvious and natural correlations between all these things as a system of symbols and facts as a possible path to two-way ET communication.

For a most basic comparison, an analogy can be made between the representation of dimensional energy from our dimension circle diagrams and the actual energy of a drop of water in a pond.

[8] "Innovations in Sound and Cymatics Advance: The Medicine of the Future," Cosmic Web, August 28, 2021.

It does not take lot of time or require much distance for the energy of the fallen waterdrop to fully dissipate into its immediate universe, that is, the pond. The physical manifestation of the rings goes away, and at that point it is no longer unique. It becomes invisible and linked to its universe as a whole. But what if there is more than one drop and the rings collide? In essence, these colliding rings are communicating with each other.

The same can be said of humans reaching the fifth dimension beyond their physical reach. If we get that far, then another entity would have to be out there to collide with in order for communication to happen at that level.

This might seem to be getting off track with respect to clues to an extraterrestrial communication code, but in fact, it is right on target. ET communication is going to be performed either up close and personal or via some sort of dimensional connection thing, or some combination of both. Many believe both forms of contact have been occurring for centuries. We are trying to get to the two-way communication part and past the one-way contact part.

It is believed that angels communicate with humans in that green space between the fourth dimension and the limit of the fifth dimension. Archangel Metatron's purpose and the sacred geometry of his cube, whose origins are closely linked to the Bible, angels, the Star of David, and sacred Jewish texts, is also linked to mystics, which is a strong message linking the 2, 3, 5, and 7 code numbers that are driving this search for clues.

The geometric layouts presented in *Angel Communication Code* so far are quite similar to something called the New Jerusalem diagram, also known as the Cosmological Circle, shown below:

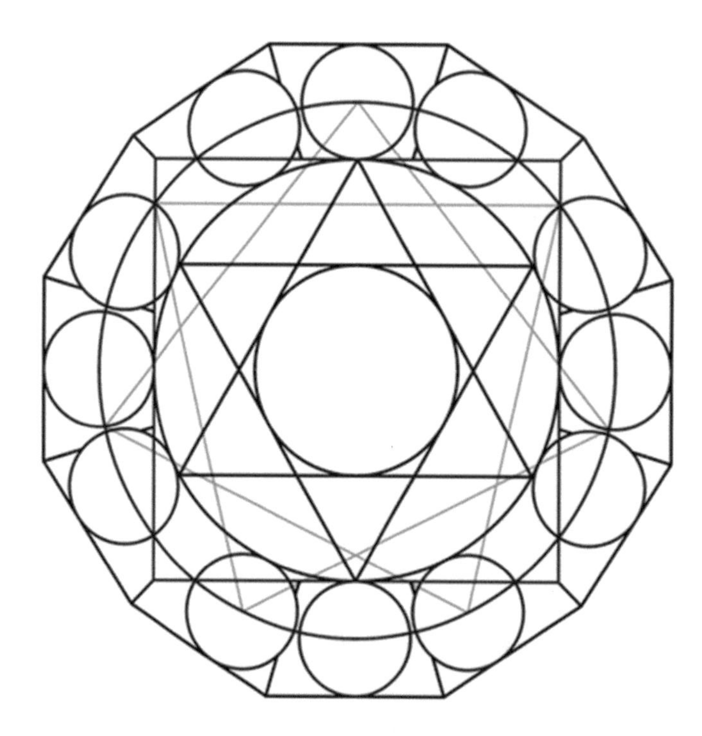

The New Jerusalem diagram is the name given by the ancient scholar John Michell. The original explanations are in John Michell's books, especially *The Dimensions of Paradise*, which describe a geometrical construction that allows the "squaring of the circle."

John Michell reconstructed the geometric pattern of the "heavenly city," which is the template of the New Jerusalem of Revelation 21. "The city was laid out as a square, its length equal to its breadth. The angel measured the city with a rod and it was 12,000 stadia in length and breadth and height. He measured its wall, and this was 144 cubits high, according to the human measure he used" (Revelation 21:16–17).

Michell discovered that this diagram served as the blueprint for many sacred sites, including Stonehenge, the Great Pyramid, St. Mary's Chapel at Glastonbury, and the city of Magnesia in Plato's *The Laws*. It also matches up with the blueprint for the Dome of the Rock, which is a shrine located on the Temple Mount in the Old City of Jerusalem. In addition to that, the Nazca, Peru, glyph called "Mandala," a.k.a. "Sun–Star–Cross," perfectly matches the New Jerusalem diagram.

Modern science confirms what our ancestors already new: the architecture of the universe (nature) is based on some key numeric proportions we can see on all levels of scale, from the atoms and DNA, through patterns and blueprints of life, all the way to solar systems and galaxies …

The Great Pyramid of Egypt, the Earth, and the Moon perfectly align and fit the New Jerusalem diagram.[9] That cannot be a fantastic coincidence. It must surely be by intelligent design. The purpose of the design is to facilitate communication with the universe.

---

9 "The Great Design," World Mysteries, May 25, 2012.

CHAPTER 7

Communication Code of Angels

It is reasonable now to reassess our historically unsuccessful efforts to send and receive messages to and from ETs. Some of the pertinent facts are as follows:

1. It is reasonable to expect that a clue or message was left for us to find that will tell us either where ETs came from or where they went and how to communicate with them. It is built into our DNA and is part of our history as humans. It is what we have done here on earth every time when new worlds were discovered. We are in fact doing it now by planting a flag on the moon and launching messages into the universe of our own design using prime numbers and binary code.
2. It is reasonable that on a primal, instinctual level, ETs would behave in a similar way to humans with respect to leaving indicators of their presence here in this new world. The evidence is everywhere we look. The problem is that we have failed to receive their message, understand it, and figure out what to do with it. Such an ET or angel message is not likely going to be a map/drawing or in some other direct hard media form.

Sending messages driven by radio telescope or satellite through space to ETs that take hundreds of light-years to be received and returned is not realistic. The communication must be sent and received in real-life time to have any meaning or value. To do that, we need to find either a portal or a way to connect to some sort of dimension of consciousness that can turn light-years into hours or days or some other reasonable amount of time.

The message we send must link as a direct response to the message that was left for us to find and be relatively simple. Launching original and overly complex messages greatly reduces the probability that if they are ever found, they will somehow be understood.

As noted several times in *Angel Communication Code*, it is reasonable to expect that any message or code left behind for us to find and decipher will be relatively simple and be built around the first three prime numbers and binary code. It is furthermore reasonable to expect the circle and triangle to be involved.

The means and methods to decipher the code discussed in *Angel Communication Code* came to the author in somewhat of a dreamlike vision in the night. That vision, combined with all the facts cited herein and all the links made to our day-to-day life and to the prime numbers 2, 3, and 5, is how we got to this point. It can also be described as my having been given a vision of a set of instructions to follow. The description of this vision is taken from my first book on this subject, *Extraterrestrial Communication Code*:

> One evening, I drifted off to sleep with that exact question burning in my head. Seemingly out of nowhere, simple numerical patterns appeared to me in some sort of dream or vision. I know what a dream is, and this was different. I woke up, remembered the patterns, and wrote them down. I then wrote a series of steps or a procedure to follow using those patterns. It popped into my head in vivid color and detail. At that moment, the exact and entire path was unclear, but the first steps down that path were very clear. Days later, I looked at what I had jotted down and felt the need to take the first steps.

This was not some sort of one-off psychotic episode or a unique experience. It has happened to many others as cited previously. Angels have always appeared before humans in dreams and visions. Dreams and visions have been responsible for some of the most significant and ahead-of-their-time, Nobel Prize–winning scientific discoveries. No longer dismissed by psychologists as random neuron firings or meaningless fantasies, dreams and visions are now considered an ongoing thought process that just happens to occur while we are asleep.[1]

Let us look at a few examples, of which there are hundreds (examples taken directly from *World of Lucid Dreaming*, "10 Dreams That Changed Human History" by Rebecca Casale).

Example 1: Niels Bohr—the Structure of the Atom

The father of quantum mechanics, Niels Bohr, often spoke of the inspirational dream that led to his discovery of the structure of the atom. Bohr was later awarded a Nobel Prize for Physics because of this leap in creative thinking in a "vision" while asleep.

The son of academic parents, Bohr got his doctorate in 1911 and gained notoriety for deciphering complex problems in the world of physics that had left his colleagues dumbfounded. He set his sights on understanding the structure of the atom, but none of his configurations would fit the contemporary model(s). One night he went to sleep and began dreaming about atoms. He saw the nucleus of the atom with electrons spinning around it much as planets spin around their sun.

Immediately upon awakening, Bohr felt inspired and believed that his vision was accurate. But as a scientist, he knew the importance of validating his idea before announcing it to the world. He returned to his lab and searched for evidence to support his theory. His theory envisioned in a dreamlike vision held true, and his vision of atomic structure turned out to be one of the greatest breakthroughs of his day.

Example 2: Albert Einstein—the Speed of Light

Albert Einstein is famous for his genius and his insights into the nature of the universe. He was awarded the Nobel Prize in Physics in 1922 for his contributions to theoretical physics.

As it happens, he came to his extraordinary scientific achievement, discovering the principle of relativity, after having a vivid vision about the speed of light that proved to be critical to his theory of relativity, which is energy (E) equals mass (m) times the speed of light (c) squared, or $E = mc^2$.

---

1 Rebecca Casale, "10 Dreams That Changed Human History," World of Lucid Dreaming, accessed November 21, 2023.

As a young man, Einstein envisioned in a dream that he was sledding down a steep mountainside, going so fast that eventually he approached the speed of light. At this moment, the stars in his dream changed their appearance in relation to him. He awoke and meditated on this idea, soon formulating what would become one of the most famous scientific theories in the history of humankind.

Example 3: Srinivasa Ramanujan—the Man Who Knew Infinity

Ramanujan shared the 1983 Nobel Prize in Physics with William A. Fowler for "theoretical studies of the physical processes of importance to the structure and evolution of the stars."

Ramanujan, a self-taught mathematical genius and prodigy from India, made substantial contributions to the analytical theory of numbers, elliptical functions, continued fractions, and infinite series. He proved more than three thousand mathematical theorems in his lifetime. Ramanujan stated that the insight for his work came to him in his visions and dreams on many occasions.

He also said that throughout his life he repeatedly dreamed of a Hindu goddess known as Namakkal or Hanuman. Lord Hanuman, known and celebrated for his devotion to Lord Rama, is the son of Anjana and Kesari and is blessed by the god of wind, Vayu. It is believed that Hanuman is an avatar of Lord Shiva or Rudra Avatar. Several texts present him as an incarnation of Lord Shiva.

Hanuman presented Ramanujan with complex mathematical formulas via dreams and visions on many occasions, which the latter could then test and verify upon waking. One such example is the infinite series for pi. Describing one of his many insightful math dreams, Ramanujan said: "While asleep I had an unusual experience. There was a red screen formed by flowing blood as it were. I was observing it. Suddenly a hand began to write on the screen. I became all attention. That hand wrote a number of results in elliptic integrals. They stuck to my mind. As soon as I woke up, I committed them to writing." There is a fascinating story behind this prolific dreamer.

From the temples and slums of Madras to the courts and chapels of Cambridge University, Ramanujan's story is one of his unlikely rise from underprivileged obscurity to the position of one of the greatest mathematical geniuses of the twentieth century.

Example 4: Otto Loewi—Nerve Impulse Breakthrough

Otto Loewi was a German-born pharmacologist whose discovery of acetylcholine (a neurotransmitter that promotes dreaming) helped advance medical therapy. He was awarded the Nobel Prize in Physiology or Medicine in 1936 jointly with Sir Henry Hallett Dale for their discoveries relating to chemical transmission of nerve impulses. Loewi is almost as famous for the means by which he discovered it as he is for the discovery itself.

In 1921, Loewi dreamed of an experiment that would prove once and for all that transmission of nerve impulses was chemical, not electrical. He woke up, scribbled the experiment down, and went back to sleep. The next morning, he arose excited to try his experiment but was horrified to find he couldn't read his midnight ramblings. That day, he said, was the longest day of his life as he tried but failed to recall his dream. The following night he had the same dream repeat itself, and upon awakening he went directly to his lab to prove the Noble Prize–winning theory of chemical transmission of the nervous impulse.

Example 5: August Kekulé—the Ouroboros Benzene Dream

A prominent German organic chemist, August Kekulé insightfully dreamed of the structure of the benzene molecule, which unlike other known organic compounds has a circular structure rather than a linear one. Kekulé died in 1896, several years before the Nobel Prizes were first awarded, but his students Jacobus H. van 't Hoff and Emil Fischer were the first two Nobel laureates in Chemistry in 1901 and 1902, respectively.

This new understanding of all aromatic compounds proved to be so important for both pure and applied chemistry after 1865 that the German Chemical Society organized an elaborate event in Kekulé's honor, where he described the dream that inspired the breakthrough.

Kekulé said he discovered the ring shape of the benzene molecule after having a vision of a snake seizing its own tail, which is a common ancient symbol known as the ouroboros, an emblematic serpent of ancient Egypt and Greece represented with its tail in its mouth, continually devouring itself and being reborn as itself. As a gnostic and alchemical symbol, the ouroboro expresses the unity of all things material and spiritual, which never disappear but perpetually change form in an eternal cycle of destruction and re-creation.[2]

Kekulé is quoted as saying:

> I was sitting writing at my textbook but the work did not progress; my thoughts were elsewhere. I turned my chair to the fire and dozed. Again the atoms were gamboling before my eyes. This time the smaller groups kept modestly in the background. My mental eye, rendered more acute by the repeated visions of the kind, could now distinguish larger structures of manifold confirmation: long rows, sometimes more closely fitted together all twining and twisting in snake like motion.
>
> But look! What was that? One of the snakes had seized hold of its own tail, and the form whirled mockingly before my eyes. As if by a flash of lightning I awoke; and this time also I spent the rest of the night in working out the rest of the hypothesis.

Example 6: Frederick Banting—Advances in Medicine

After his mother passed away from diabetes, Frederick Banting was motivated to find a cure. Eventually he found the next best thing: a treatment using insulin injections, which, though not a true cure, could at least significantly extend the life span of the afflicted. The discovery won him a Nobel Prize in Medicine in 1923 at just thirty-two years old.

Although he lacked knowledge of diabetes and clinical research, Banting's unique knowledge of surgery combined with his assistant's (Charles Best's) knowledge of diabetes led the two men to become the ideal research team. While seeking to isolate the exact cause of diabetes, Banting had a dream telling him to surgically ligate (tie up) the pancreas of a diabetic dog in order to stop the flow of nourishment. He did so, and discovered a disproportionate balance between sugar and insulin. This breakthrough led to another dream that revealed how to develop insulin as a drug to treat the condition.

---

2 *Encyclopedia Britannica*, s.v. "Ouroboros."

Banting was named Canada's first professor of medical research, and by 1923 he was the most famous man in the country. He received letters and gifts from hundreds of grateful diabetics all over the world, and since then insulin has saved or transformed the lives of millions of people.

Until a person has the experience himself, he has to be skeptical about these sorts of dreamlike visions. However, there are many examples of very credible people having such visions throughout history. We are not talking about angels coming down from heaven to speak to humans, or God Himself doing the same. We are talking about simple visions seen or heard in dreams or at some other level of consciousness. This author never anticipated that it would happen to him, but it did. The procedure was quite clear in the vision/dream, but it was only the first step. The trail had to be followed, which leads us here today. The justification and defense for this journey is what *Angel Communication Code* demonstrates, leading to the hypothesis and ultimately the experiment to put it all to the test.

The *Extraterrestrial Communication Code* book did not dig as deeply into the idea of angels or into biblical references to messengers revealing an ET communication code as *Angel Communication Code* does. Thus far we should be able to agree that there are many conspicuous occurrences of the numbers 2, 3, 5, and 7 linked to the story of angels, with the number 3 arguably being the most prominent.

In the Bible, the number 7 is often used as symbol of perfection or completion. Genesis, the very first book of the Old Testament, explains that God created the heavens and the earth in six days and then rested on the seventh day (Genesis 1, 2:1–2). The number 7 is especially prominent in scripture, appearing more than seven hundred times from the first book of the Bible, Genesis, explaining the creation of the universe, to the last book of the Bible, Revelation, describing the end of the world, and in many places in between. The number 7 is used to identify concepts such as exoneration, healing, completion, perfection, and the fulfillment of promises and prayers. Let us look at some of these important correlations:

Exoneration and Healing

1. Deuteronomy (15:1–2, 12) says that every seventh year, the Israelites were to cancel all the debts they had made with each other and free their slaves. They were to be exonerated.
2. In the book of Matthew, the number 7 is connected with exoneration when Peter asks Jesus how many times we are to forgive each other and Jesus replies, "seventy times seven" times (Matthew 18:21–22). In that instance, Jesus isn't telling us to literally forgive someone 490 times. Christ is instructing us to forgive each other wholly, as His Jewish audience would have understood. The emphasis on the number 7 in this teaching is to indicate complete forgiveness. Why use the number 7? Because it has another meaning than the obvious message being delivered.
3. In the context of healing, the prophet Elisha referenced the number 7 when he directed Naaman the leper to bathe in the Jordan River seven times to be healed (2 Kings 5:9–10, 14). Why not any other number of times but seven? Because it has another meaning than the obvious message being delivered.
4. We see a link between seven and healing in the seven, and only seven, "healing" miracles that Jesus, the future King of all angels, performed on the Sabbath, or the seventh day of the week. Specifically, Jesus healed the following seven people on the Sabbath day:
    - a man with a deformed hand (Matthew 12:9–13);

- a man possessed by an unclean spirit (Mark 1:23–26);
- Peter's mother-in-law with fever (Mark 1:29–31);
- a woman crippled by a spirit (Luke 13:10–13);
- a man with abnormal swelling of the body (Luke 14:1–4);
- a lame man by the pool of Bethesda (John 5:5–9); and
- a man born blind (John 9:1–7).

Completion and Perfection

1. As previously noted, the book of Genesis tells us that God created the heavens and the earth in six days and, upon completion, rested on the seventh day (Genesis 1, 2:1–2). Based on this cycle of work and rest, God commands us to also labor for six days and then complete the week by resting on the seventh day, the day God set apart as the holy Sabbath (Exodus 20:9–11).
2. The number 7 also denotes completion at the Crucifixion, when Jesus spoke seven statements in agony from the cross at the completion of His earthly duties:
   - "Father, forgive them, for they know not what they do" (Luke 23:34).
   - "Truly, I say to you, today you will be with Me in Paradise" (Luke 23:43).
   - "'Woman, behold, your son!' Then He said to the disciple, 'Behold, your mother!'" (John 19:26–27).
   - "My God, My God, why have you forsaken Me?" (Matthew 27:46).
   - "I thirst" (John 19:28).
   - "It is finished" (John 19:30).
   - "Father, into Your hands I commit My spirit!" (Luke 23:46).

In the context of perfection, Jesus spoke in a grouping of seven when He was asked how we should pray (Matthew 6:9–13). In response, He gave us the Lord's Prayer, surely a perfect way to pray considering that the words came from Christ Himself. Notably, the Lord's Prayer contains seven petitions:

1. "hallowed be Thy name";
2. "Thy kingdom come";
3. "Thy will be done on earth as it is in heaven";
4. "give us this day our daily bread";
5. "forgive us our trespasses as we forgive those who trespass against us";
6. "lead us not into temptation"; and
7. "deliver us from evil."

Jesus again spoke in a grouping of seven when He used seven metaphors to describe Himself as the path to salvation, the perfect reward for a good and faithful servant. Jesus tells us He is:

1. the bread of life (John 6:35);
2. the light of the world (John 8:12);
3. the gate to salvation (John 10:9);
4. the Good Shepherd (John 10:11);
5. the resurrection and the life (John 11:25–26);
6. the way, the truth, and the life (John 14:6); and
7. the Vine (John 15:5).

King David referenced the number seven when describing the perfect nature of God's Words when he wrote that the Lord's Words are flawless, "like gold refined seven times" (Psalm 12:6). Recall all the other unmistakable relationships between the Star of David, Metatron's Cube, and the Vitruvian Man previously brought together in *Angel Communication Code.*

When the prophet Isaiah described the coming Messiah, he listed seven qualities that the Savior would embody (Isaiah 11:1–2).

Fulfillment of Promises and Oaths

The number 7 also frequently accompanies the fulfillment of promises or oaths. In fact, the Hebrew word for swearing an oath (*Shaba*) and the Hebrew word for seven (*Sheba*) both derive from the Hebrew word meaning satisfaction or fullness (*Saba*).

In Genesis, God promises not to destroy the earth again with a flood and memorializes this covenant with the rainbow, which is made up of seven colors (Genesis 9:8–15). Later in Genesis, we learn that Abraham swore an oath of ownership over a certain well of water (Genesis 21:22–31). He satisfied the oath with a gift of seven lambs and named the site of the oath "Beersheba," which interchangeably means "well of the oath" or "well of seven."

The book of Joshua gives us another example of the number 7's correlation with promises. God promised Joshua that He would bring down the fortified walls of Jericho if Joshua and his army marched around the city once for six days and seven times on the seventh day with seven priests blowing seven trumpets. After Joshua followed the Lord's commands exactly, the walls of Jericho fell, just as the Lord had promised (Joshua 6:1–20).

Seven is also associated with promises in the book of Revelation. In particular, we read of seven letters addressed to seven churches (Revelation 2–3). In the letters, Christ assures each church community that if its members repent and live according to His instructions, each community will receive its promised reward.

In addition, Revelation often invokes the number 7 in its discussion of God's promise to save those whose names are written in the book of life and condemn those whose names aren't (Revelation 20:15, 21:1, 21:27). The fulfillment of this divine promise is ushered in by groups of seven: seven seals, seven trumpets sounded by seven angels, and seven bowls of God's promised wrath carried by seven angels (Revelation 6, 8, 11, 16).

The Bible's extensive use of the number 7 in connection with such concepts as completion, exoneration, and the fulfillment of promises suggests that God ascribes a sacred nature to the number. However, we must balance this conclusion with the fact that not every mention of the number 7 in scripture has an alternate underlying implication. When we scrutinize the Bible and angels, we can find a host of numbers that have deeper meanings with threads throughout the passages. These show the intricate handiwork of God's plan and the amazing ways it unfolds throughout history.[3]

It is important again to point out and recognize that the Bible is not the Word of God as written or dictated by God. It is not a biography, nor is it intended to be. It is a collection of books authored by approximately forty authors over the course of approximately fifteen hundred years, with some of the authors unknown.

---

3 Dolores Smyth, "What Is the Biblical Significance of the Number 7?," Christianity.com, January, 2020.

This fact opens the door to the additional possibility that these authors may have had a secondary agenda of close-to-equal importance as the importance of the primary message. That secondary agenda may have been the inclusion of a coded cipher for future discovery that leads us off Earth and out into the universe of God's other intelligent creatures.

The definition of the word *intelligence* is very complex. We arrogantly call ourselves intelligent, but intelligent compared to what or whom? Perhaps we need to demonstrate our so-called intelligence to higher-level beings other than ourselves. We must not allow our blind arrogance and self-proclaimed intelligence and importance to result in our downfall. The Bible has many verses that discuss humility before God.

It is a common belief that ETs have been influencing us and watching over us for many centuries. Their purpose, however, remains a mystery. Now is the time to establish direct two-way communication with them by finding and using the tools provided to us by them and to learn their purpose.

It is important to understand that it is not the intent of *Angel Communication Code* to twist, misrepresent, or discredit the Holy Scripture's message about God and all that goes with the angels and the Christian faith. It is not the intent of *Angel Communication Code* to discredit the scientists who have tried and thus far failed to achieve ET communication. It is the intent of *Angel Communication Code* to inspire a different way of thinking and searching for that ET message. The intent is to demonstrate that there is another important message buried within the holy texts and the messages of angels. Those messages are in the entirety of the Bible and, more specifically, part of the concept of angels and the structure of angels defined by humans. The clues are in the frequent and conspicuous use of prime numbers 2, 3, 5, and 7. The more than forty authors of the books that make up the Bible over a course of fifteen hundred years provided the means, motive, and opportunity for such an important message, all the things necessary to warrant a criminal trial in a court of law.

Many messages have been transmitted from Earth out into the universe over the years, mostly via radio signals. A radio communication system includes a radio transmitter, a free space communication channel, and a radio receiver. At the top level, a radio transmitter system consists of a data interface, a modulator, a power amplifier, and an antenna. The transmitter system uses the modulator to encode digital data onto a high-frequency electromagnetic wave. The power amplifier then increases the output radio frequency (RF) power of the transmitted signal to be sent through free space to the receiver using the transmit antenna. The radio receiver system uses a receiving antenna, low-noise amplifier, and demodulator to produce digital data output from the received signal. The receiving antenna collects the electromagnetic waves and routes the signal to the receiver, which then demodulates the wave and converts the electrical signals back into the original digital message. Low-noise amplifiers and filters are sometimes employed to reduce signal noise in certain frequency bands or to increase the received signal strength. In many cases, the functions of the modulator and demodulator are combined into a radio transceiver that can both send and receive RF signals.[4]

There are twelve recognized radio space message "projects" from the USA, as follows:[5]

1. the Morse Message (1962)
2. the Arecibo message (1974)

---

4 NASA, "State-of-the-Art Small Spacecraft Technology," https://www.nasa.gov/smallsat-institute/sst-soa/, 2021.

5 Wikipedia, s.v. "List of Interstellar Radio Messages."

3. Cosmic Call 1 (1999), four transmissions to nearby sunlike stars
4. The Teen Age Message (2001), six transmissions
5. Cosmic Call (2003), five transmissions
6. A Message from Earth (2008), one transmission to the Gliese 581 planetary system
7. Across the Universe (2008)
8. Hello from Earth (2009), one transmission to the Gliese 581 planetary system
9. Wow! Reply (2012), three transmissions to Hipparcos 34511, Hipparcos 33277, and Hipparcos 43587 in reply to the Wow! Signal
10. Lone Signal (2013)
11. A Simple Response to an Elemental Message (2016)
12. Sónar Calling (2017).

In addition to radio messages, there are five interstellar probes, all launched by NASA:[6]

1. *Pioneer 10* (1972)
2. *Pioneer 11* (1973)
3. *Voyager 1* (1977)
4. *Voyager 2* (1977)
5. *New Horizons* (2006).

As of 2019, *Voyager 1*, *Voyager 2*, and *Pioneer 10* are the only probes to have reached interstellar space. The other two remain on interstellar trajectories. All are tracked by radio signals.

None of the fifteen projects have produced any credible results with respect to extraterrestrial communication as far as the general public is aware to date. These projects were designed and put together by some of the smartest scientists and engineers on the face of the planet, so the question is, what is (or what is not) going on here?

Scientists are now beginning to acknowledge the shortcomings of our efforts to date and are rethinking the process as the scientific method demands. We are also doing the same in *Angel Communication Code*, but in a different and unique way. NASA's current thinking on this subject goes like this (taken from Daniel Oberhaus, "Researchers Made a New Message for Extraterrestrials," *Scientific American*, March 30, 2022):

> Upon discovering the existence of intelligent life beyond Earth, the first question we are most likely to ask is "How can we communicate?" As we approach the fiftieth anniversary of the 1974 Arecibo message—humanity's first attempt to send out a missive capable of being understood by extraterrestrial intelligence—the question feels more urgent than ever. Advances in remote sensing technologies have revealed that the vast majority of stars in our galaxy host planets and that many of these exoplanets appear capable of hosting liquid water on their surface—a prerequisite for life as we know it. The odds that at least one of these billions of planets has produced intelligent life seem favorable enough to spend some time figuring out how to say hello.
>
> In early March, an international team of researchers led by Jonathan Jiang of NASA's Jet Propulsion Laboratory posted a paper on the preprint server arXiv.org that detailed a new

[6] Wikipedia, s.v. "Interstellar Probe."

design for a message intended for extraterrestrial recipients. The thirteen-page epistle, referred to as the "Beacon in the Galaxy," is meant to be a basic introduction to mathematics, chemistry, and biology that draws heavily on the design of the Arecibo message and other past attempts at contacting extraterrestrials. The researchers included a detailed plan for the best time of year to broadcast the message and proposed a dense ring of stars near the center of our galaxy as a promising destination. Importantly, the transmission also features a freshly designed return address that will help any alien listeners pinpoint our location in the galaxy so they can—hopefully—kick off an interstellar conversation.

"The motivation for the design was to deliver the maximum amount of information about our society and the human species in the minimal amount of message," Jiang says. "With improvements in digital technology, we can do much better than the [Arecibo message] in 1974."

Every interstellar message must address two fundamental questions: what to say and how to say it. Nearly all the messages that humans have broadcast into space so far start by establishing common ground with a basic lesson in science and mathematics, two topics that are presumably familiar to both [us] and to extraterrestrials. If a civilization beyond our planet is capable of building a radio telescope to receive our message, it probably knows a thing or two about physics. A far messier question is how to encode these concepts into the communiqué. Human languages are out of the question for obvious reasons, but so are our numeral systems. Though the concept of numbers is nearly universal, the way we depict them as numerals is entirely arbitrary. This is why many attempts, including "Beacon in the Galaxy," opt to design their letter as a bitmap, a way to use binary code to create a pixelated image.

The bitmap design philosophy for interstellar communication stretches back to the Arecibo message. It is a logical approach—the on/off, present/absent nature of a binary seems like it would be recognized by any intelligent species. But the strategy is not without its shortcomings. When pioneering search for extraterrestrial intelligence (SETI) scientist Frank Drake designed a prototype of the Arecibo message, he sent the binary message by post to some colleagues, including several Nobel laureates. None of them were able to understand its contents, and only one figured out that the binary was meant to be a bitmap. If some of the smartest humans struggle to understand this form of encoding a message, it seems unlikely that an extraterrestrial would fare any better. Furthermore, it is not even clear that space aliens will be able to see the images contained within the message if they do receive it.

"One of the key ideas is that, because vision has evolved independently many times on Earth, that means aliens will have it, too," says Douglas Vakoch, president of METI (Messaging Extraterrestrial Intelligence) International, a nonprofit devoted to researching how to communicate with other life-forms. "But that's a big 'if,' and even if they can see, there is so much culture embedded in the way we represent objects. Does that mean we should rule out pictures? Absolutely does not. It means we should not naively assume that our representations are going to be intelligible."

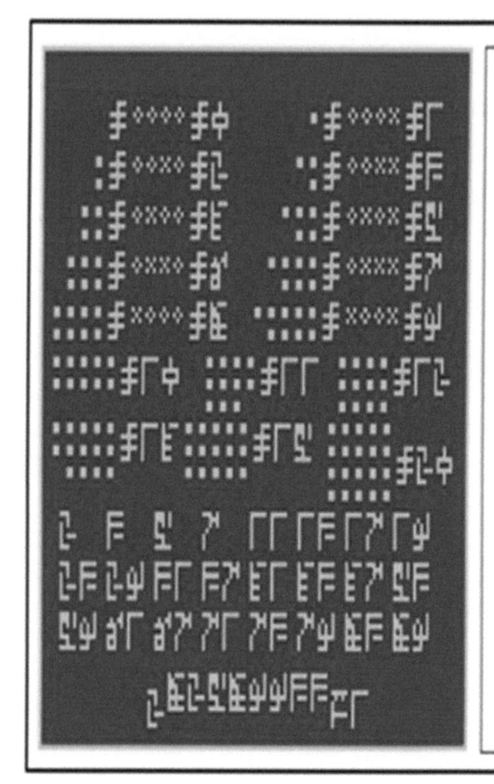

Binary and Decimal Systems

0 dot = 0000 = "0"	1 dot = 0001 = "1"
2 dots = 0010 = "2"	3 dots = 0011 = "3"
4 dots = 0100 = "4"	5 dots = 0101 = "5"
6 dots = 0110 = "6"	7 dots = 0111 = "7"
8 dots = 1000 = "8"	9 dots = 1001 = "9"

10 dots = "10" 11 dots = "11" 12 dots = "12"
14 dots = "14" 15 dots = "15" 20 dots = "20"

Prime Numbers

2, 3, 5, 7, 11, 13, 17, 19, 23, 29, 31, 37, 1, 43, 47, 53, 59, 61, 67, 71, 73, 79, 83, 89

The largest known Prime Number

$2^{82,589,933}-1$

In 2017 Vakoch and his colleagues sent the first interstellar message transmitting scientific information since 2003 to a nearby star. It, too, was coded in binary, but it eschewed bitmaps for a message design that explored the concepts of time and radio waves by referring back to the radio wave carrying the message. Jiang and his colleagues chose another path. They based much of their design on the 2003 Cosmic Call broadcast from the Yevpatoriaradio telescope in the region of Crimea. This message featured a custom bitmap "alphabet" created by physicists Yvan Dutil and Stéphane Dumas as a protoalien language that was designed to be robust against transmission errors.

After an initial transmission of a prime number to mark the message as artificial, Jiang's message uses the same alien alphabet to introduce our base-10 numeral system and basic mathematics. With this foundation in place, the message uses the spin-flip transition of a hydrogen atom to explain the idea of time and mark when the transmission was sent from Earth, introduce common elements from the periodic table, and reveal the structure and chemistry of DNA. The final pages are probably the most interesting to extraterrestrials but also the least likely to be understood because they assume that the recipient represents objects in the same way that humans do. These pages feature a sketch of a male and female human, a map of Earth's surface, a diagram of our solar system, the radio frequency that the extraterrestrials should use to respond to the message, and the coordinates of our solar system in the galaxy referenced to the location of globular clusters—stable and tightly packed groups of thousands of stars that would likely be familiar to an extraterrestrial anywhere in the galaxy.

"We know the location of more than fifty globular clusters," Jiang says. "If there's an advanced civilization, we bet that, if they know astrophysics, they know the globular cluster locations as well, so we can use this as a coordinate to pinpoint the location of our solar system."

Jiang and his colleagues propose sending their message from either the Allen Telescope Array in Northern California or the Five-Hundred-Meter Aperture Spherical Radio Telescope (FAST) in China. Since the recent destruction of the Arecibo telescope in Puerto Rico, these two radio telescopes are the only ones in the world that are actively courting SETI researchers. At the moment, though, both telescopes are only capable of listening to the cosmos, not talking to it. Jiang acknowledges that outfitting either telescope with the equipment required to transmit the message will not be trivial. But doing so is possible, and he says he and his co-authors are discussing ways to work with researchers at FAST to make it happen.

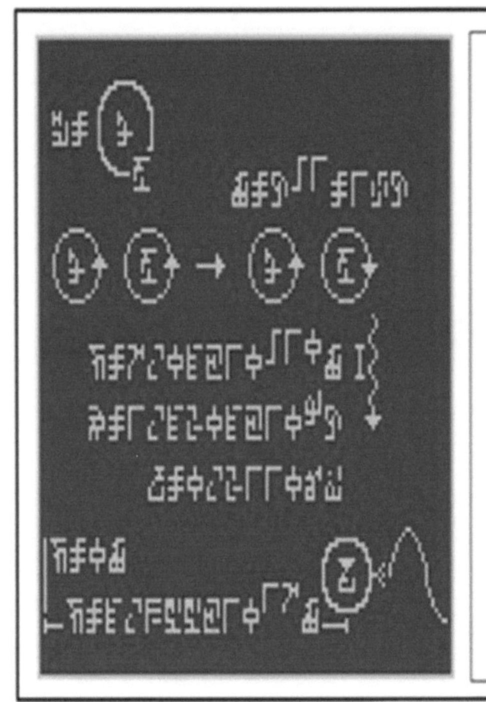

Particle Physics

H="electron (×1)" orbiting around "proton (×1)"

"second"="hertz"$^{-1}$=1/"hertz"

(Hydrogen spin-flip from unidirectional to bidirectional)

"time"=7.04×10^{-10} "second"
"frequency"=1.4204×10^{9} "hertz" (EM radiation)
"wavelength"=0.21106 "meter"

(From "time"=0 "second to message sent from Earth)
"time"=4.355×10^{17} "second"

If Jiang and his colleagues get a chance to transmit their message, they calculated that it would be best to do so sometime in March or October, when Earth is at a ninety-degree angle between the sun and its target at the center of the Milky Way. This would maximize the chance that the missive would not get lost in the background noise of our host star. But a far deeper question is whether we should be sending a message at all.

Messaging extraterrestrials has always occupied a controversial position in the broader SETI community, which is mostly focused on listening for alien transmissions rather than sending out our own. To detractors of "active SETI," the practice is a waste of time at best and an existentially dangerous gamble at worst. There are billions of targets to choose from, and the odds that we send a message to the right planet at the right time are dismally low. Plus, we have no idea who may be listening. What happens if we give our address to an alien species that lives on a diet of bipedal hominins?

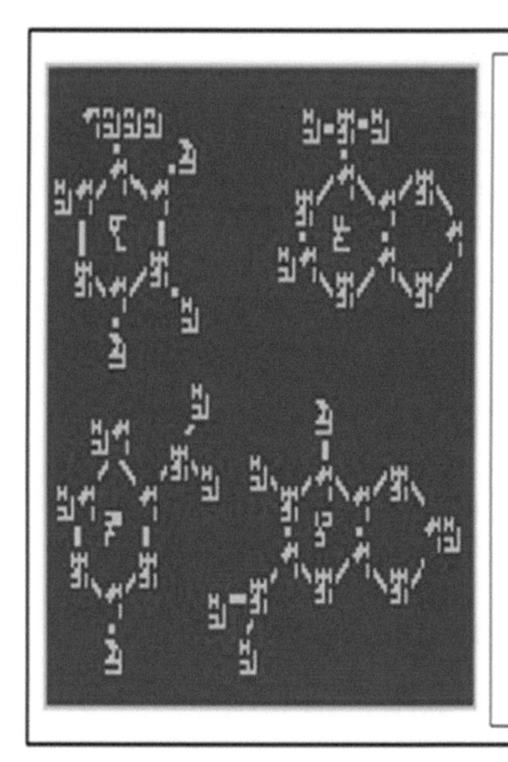

DNA Structures

Thymidine Adenosine

Cytidine Guanosine

A sample from a new message intended to be sent toward potential intelligent extraterrestrials in the galaxy, "A Beacon in the Galaxy: Updated Arecibo Message for Potential FAST and SETI Projects," by Jonathan H. Jiang et al., preprint posted online March 4, 2022 (CC BY-NC-SA 4.0), is shown below.

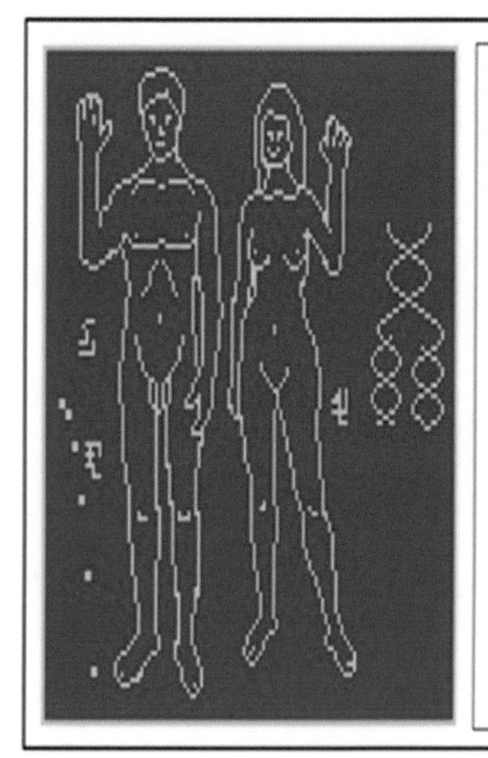

Illustrations of the Human Form

Depiction of human male and female

Object Falling Downwards for Direction (left)

Double Helix Structure (right)

"I don't live in fear of an invading horde, but other people do. And just because I don't share their fear doesn't make their concerns irrelevant," says Sheri Wells-Jensen, an associate professor of English at Bowling Green State University and an expert on the linguistic and cultural issues associated with interstellar message design. "Just because it would be difficult to achieve global consensus on what to send or whether we should send doesn't mean we shouldn't do it. It is our responsibility to struggle with this and include as many people as possible."

Despite the pitfalls, many insist that the potential rewards of active SETI far outweigh the risks. First contact would be one of the most momentous occasions in the history of our species, the argument goes, and if we just wait around for someone to call us, it may never happen. As for the risk of annihilation by a malevolent space alien: We blew our cover long ago. Any extraterrestrial capable of traveling to Earth would be more than capable of detecting evidence of life in the chemical signatures of our atmosphere or the electromagnetic radiation that has been leaking from our radios, televisions, and radar systems for the past century. "This is an invitation to all people on Earth to participate in a discussion about sending out this message," Jiang says. "We hope, by publishing this paper, we can encourage people to think about this."[7]

The NASA article raises many very good and valid points. There are considerations that seem to be absent from the article, such as:

1. There is no solution to overcoming the distance issue and the time it would take to get our message to its recipient.
2. The new message is still complicated and may not be understood if received.
3. It assumes we have not received an original message from ETs, versus seeking a response to any of the messages we have already launched.
4. The new approach is essentially the same as the failed approaches of the past but with a somewhat simplified original message. It is still a complicated message. It does not address that problem. The scientists are sniffing in the same old hole.

What is all this chatter and reference to some sort of code laid out in a vision that has a specific structure and procedure? Let us walk through it in a way that is consistent with the layout of the angel hierarchy, which involves three triads of three angel choirs laid out in the same vertical manner.

The vision/dream started very simply. Construction begins by taking the basic 2, 3, 5 input numbers and turning that into binary code. The original version started with this simple image:

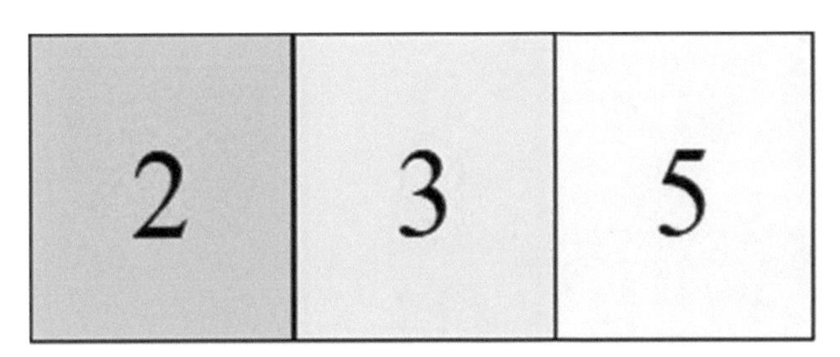

7 Daniel Oberhaus, "Researchers Made a New Message for Extraterrestrials," *Scientific American*, March 30, 2022.

Then it expanded into this:

2	3	5
3	5	2
5	2	3

Notice how the 2, 3, 5 sequences rotate both vertically and horizontally, a very basic rotational concept.

Then the geometry of the blocks changed into this more rectangular configuration:

Same pattern—different geometry

2	3	5
3	5	2
5	2	3

Now the vision subdivided the blocks above three times, and it appeared like this:

2	3	5	3	5	2	5	2	3
3	5	2	5	2	3	2	3	5
5	2	3	2	3	5	3	5	2

Notice how the 2, 3, 5 order of numbers rotates within the sequence. You can now begin to see relevant patterns of triangles.

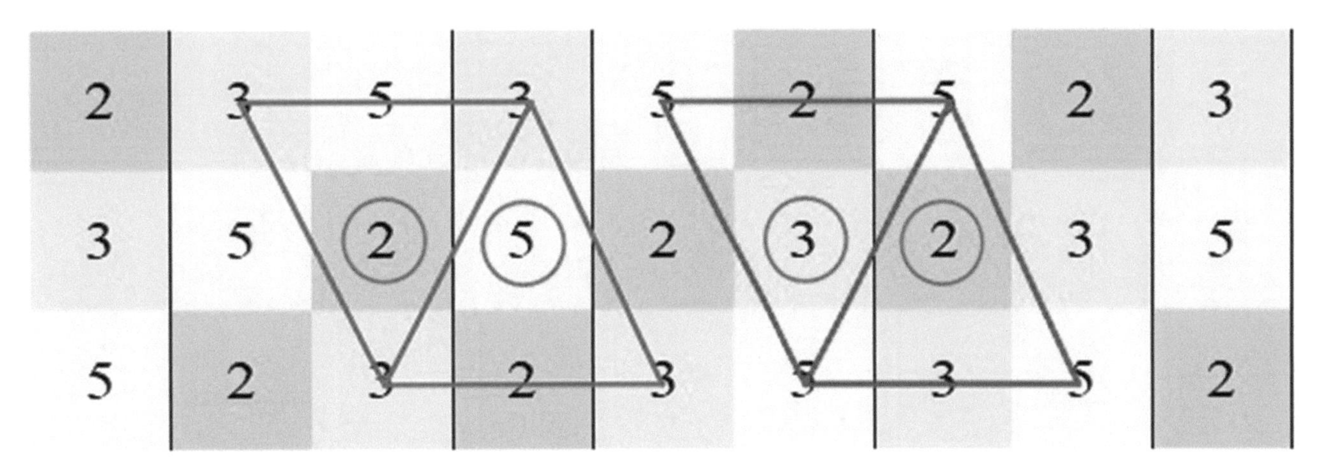

Looking more closely, we can see that:

1. The 3 and the 5 (and not the 2) both form two triangles using the top and bottom rows, one pointing up and one pointing down, as we saw in the Star of David.
2. The number 3 triangle has a 2 and a 5 within its boundaries, completing the 2, 3, 5 sequence.
3. The number 5 triangle has a 2 and a 3 within its boundaries, also completing the 2, 3, 5 sequence.
4. Both the number 3 and the number 5 triangles each hold the number 2. One is in the number 3 triangle pointing down, and the other is in the 5 triangle pointing up. This means that the center of each of the two triangles has the number 2, the first prime number, in its center.

5. There is no number 2 triangle that fits this pattern; however, it is common to both the number 3 and the number 5 triangles, one pointing up and one pointing down. This is all very much akin to the Star of David triangle when we brought two triangles together that share a common core. In the Star of David, the core represents the first dimension of consciousness. The vision core when we combine the number 3 and the number 5 triangles is the number 2, the first prime number. The vision core analysis, the Star of David core, and the analysis of the Vitruvian Man are not linked in their derivation and, because of this, represent an independent verification of the significance of this double triangle clue and its core.

The vision continued. The next thing that appeared was this (minus the row and column numbering):

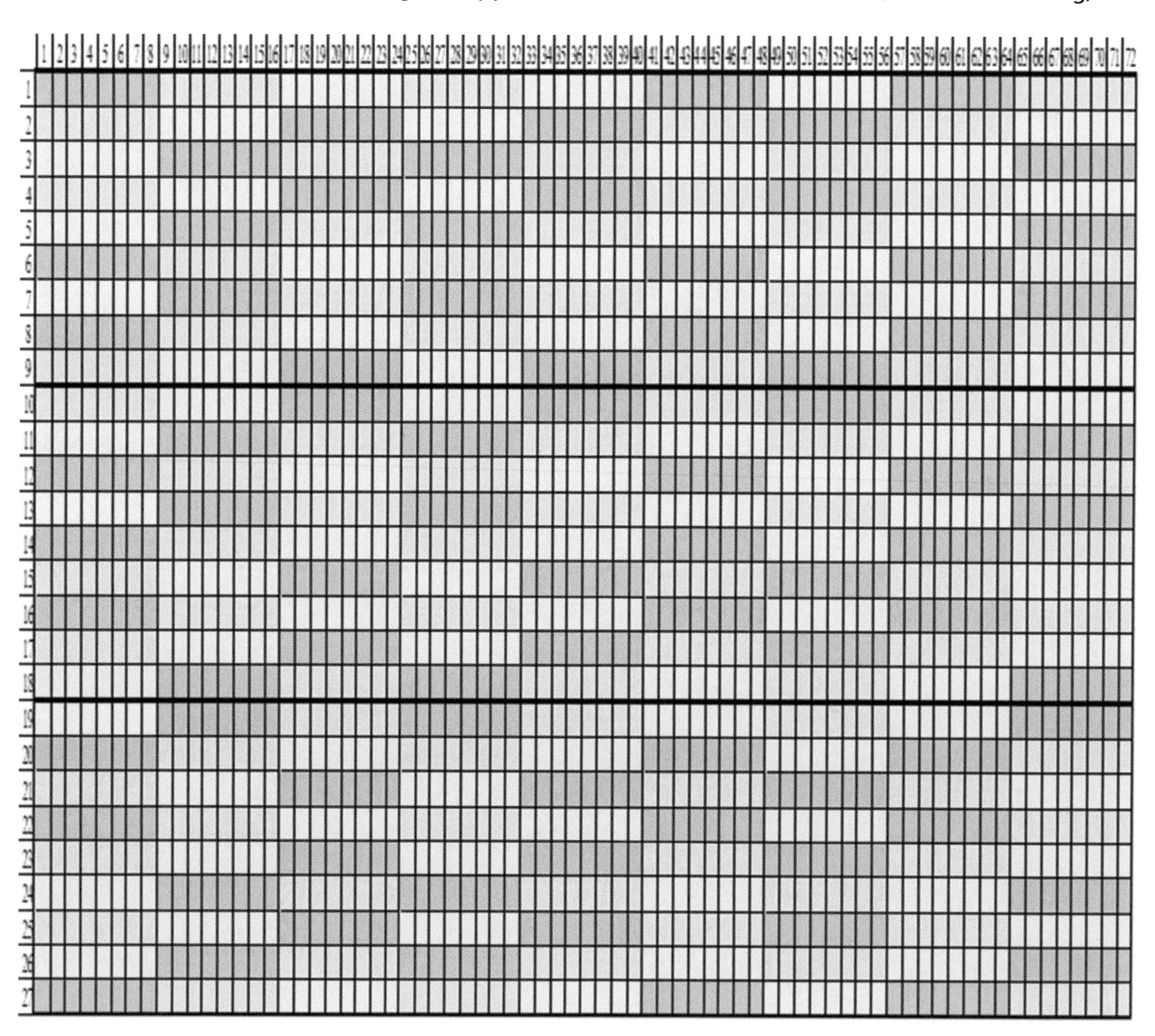

It expanded by three again then divided each block equally, resulting in $27 \times 72 = 1{,}944$ individual blocks, color coded in concert with the 2, 3, 5, colors. Recall we discussed "prime factorization" and how every number can be expressed as a product of prime numbers. Our giant block of blocks breaks down like this:

$27 = 3 \times 3 \times 3$, or 3^3

$72 = 2 \times 2 \times 2 \times 3 \times 3$, or $2^3 \times 3^2$

$1{,}944 = 2 \times 2 \times 2 \times 3 \times 3 \times 3 \times 3 \times 3$, or $2^3 \times 3^5$

It breaks down into nothing but only the first three prime numbers configured in expressions of the first three prime numbers.

The block lays out perfectly for the insertion of 2, 3, and 5 binary like this:

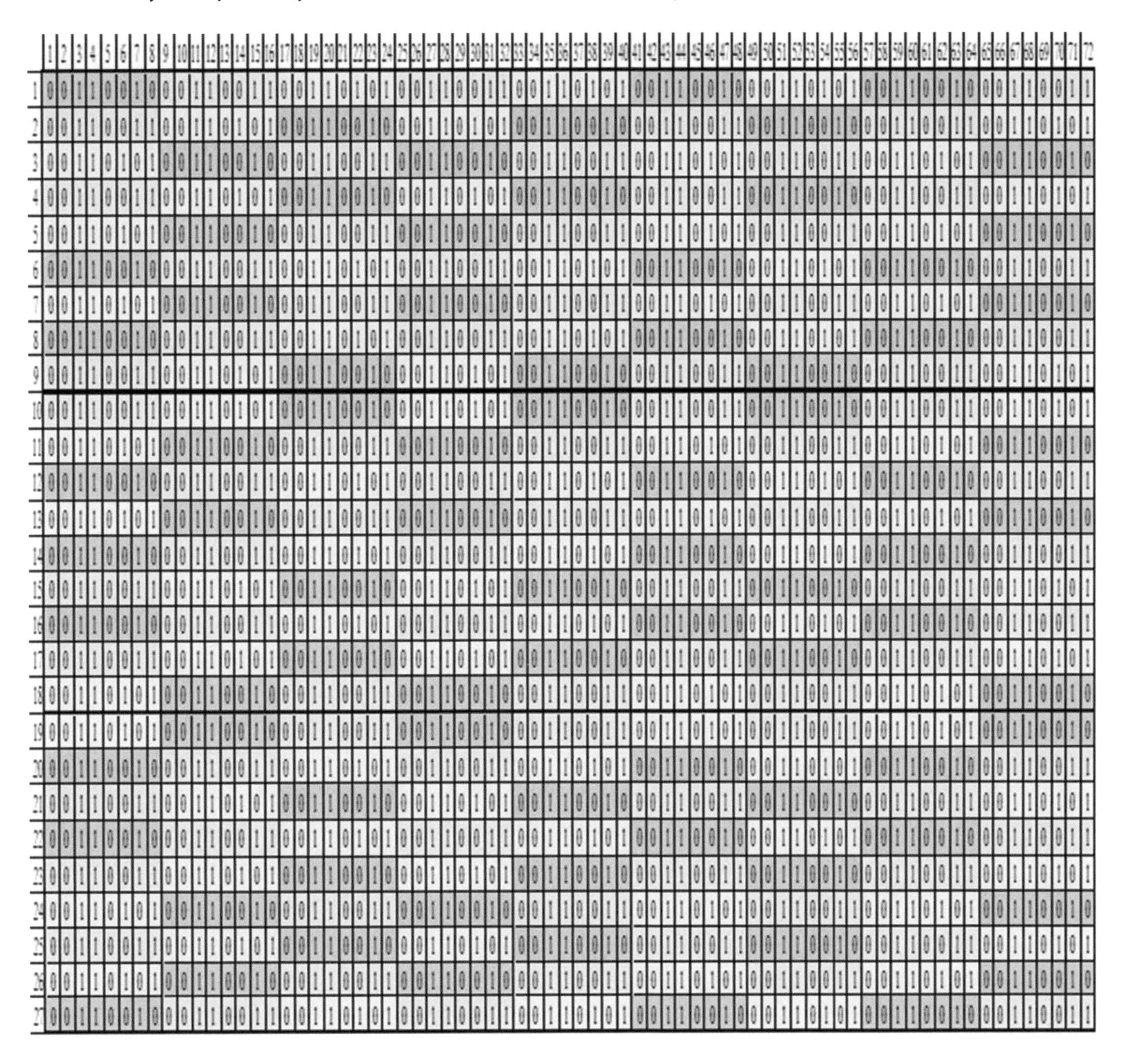

	1	2	3	4	5	6	7	8	9	10	11	12	13	14	15	16	17	18	19	20	21	22	23	24	25	26	27	28	29	30	31	32	33	34	35	36	37	38	39	40	41	42	43	44	45	46	47	48	49	50	51	52	53	54	55	56	57	58	59	60	61	62	63	64	65	66	67	68	69	70	71	72
1	0	0	1	1	0	0	1	0	0	0	1	1	0	0	1	1	0	0	1	1	0	1	0	1	0	0	1	1	0	0	1	1	0	0	1	1	0	1	0	1	0	0	1	1	0	0	1	0	0	0	1	1	0	1	0	1	0	0	1	1	0	0	1	0	0	0	1	1	0	0	1	1
2	0	0	1	1	0	0	1	1	0	0	1	1	0	1	0	1	0	0	1	1	0	0	1	0	0	0	1	1	0	1	0	1	0	0	1	1	0	0	1	0	0	0	1	1	0	0	1	1	0	0	1	1	0	0	1	0	0	0	1	1	0	0	1	1	0	0	1	1	0	1	0	1
3	0	0	1	1	0	1	0	1	0	0	1	1	0	0	1	0	0	0	1	1	0	0	1	1	0	0	1	1	0	0	1	0	0	0	1	1	0	0	1	1	0	0	1	1	0	1	0	1	0	0	1	1	0	0	1	1	0	0	1	1	0	1	0	1	0	0	1	1	0	0	1	0
4	0	0	1	1	0	0	1	1	0	0	1	1	0	1	0	1	0	0	1	1	0	0	1	0	0	0	1	1	0	1	0	1	0	0	1	1	0	0	1	0	0	0	1	1	0	0	1	1	0	0	1	1	0	0	1	0	0	0	1	1	0	0	1	1	0	0	1	1	0	1	0	1
5	0	0	1	1	0	1	0	1	0	0	1	1	0	0	1	0	0	0	1	1	0	0	1	1	0	0	1	1	0	0	1	0	0	0	1	1	0	0	1	1	0	0	1	1	0	1	0	1	0	0	1	1	0	0	1	1	0	0	1	1	0	1	0	1	0	0	1	1	0	0	1	0
6	0	0	1	1	0	0	1	0	0	0	1	1	0	0	1	1	0	0	1	1	0	1	0	1	0	0	1	1	0	0	1	1	0	0	1	1	0	1	0	1	0	0	1	1	0	0	1	0	0	0	1	1	0	1	0	1	0	0	1	1	0	0	1	0	0	0	1	1	0	0	1	1
7	0	0	1	1	0	1	0	1	0	0	1	1	0	0	1	0	0	0	1	1	0	0	1	1	0	0	1	1	0	0	1	0	0	0	1	1	0	0	1	1	0	0	1	1	0	1	0	1	0	0	1	1	0	0	1	1	0	0	1	1	0	1	0	1	0	0	1	1	0	0	1	0
8	0	0	1	1	0	0	1	0	0	0	1	1	0	0	1	1	0	0	1	1	0	1	0	1	0	0	1	1	0	0	1	1	0	0	1	1	0	1	0	1	0	0	1	1	0	0	1	0	0	0	1	1	0	1	0	1	0	0	1	1	0	0	1	0	0	0	1	1	0	0	1	1
9	0	0	1	1	0	0	1	1	0	0	1	1	0	1	0	1	0	0	1	1	0	0	1	0	0	0	1	1	0	1	0	1	0	0	1	1	0	0	1	0	0	0	1	1	0	0	1	1	0	0	1	1	0	0	1	0	0	0	1	1	0	0	1	1	0	0	1	1	0	1	0	1
10	0	0	1	1	0	0	1	1	0	0	1	1	0	1	0	1	0	0	1	1	0	0	1	0	0	0	1	1	0	1	0	1	0	0	1	1	0	0	1	0	0	0	1	1	0	0	1	1	0	0	1	1	0	0	1	0	0	0	1	1	0	0	1	1	0	0	1	1	0	1	0	1
11	0	0	1	1	0	1	0	1	0	0	1	1	0	0	1	0	0	0	1	1	0	0	1	1	0	0	1	1	0	0	1	0	0	0	1	1	0	0	1	1	0	0	1	1	0	1	0	1	0	0	1	1	0	0	1	1	0	0	1	1	0	1	0	1	0	0	1	1	0	0	1	0
12	0	0	1	1	0	0	1	0	0	0	1	1	0	0	1	1	0	0	1	1	0	1	0	1	0	0	1	1	0	0	1	1	0	0	1	1	0	1	0	1	0	0	1	1	0	0	1	0	0	0	1	1	0	1	0	1	0	0	1	1	0	0	1	0	0	0	1	1	0	0	1	1
13	0	0	1	1	0	1	0	1	0	0	1	1	0	0	1	0	0	0	1	1	0	0	1	1	0	0	1	1	0	0	1	0	0	0	1	1	0	0	1	1	0	0	1	1	0	1	0	1	0	0	1	1	0	0	1	1	0	0	1	1	0	1	0	1	0	0	1	1	0	0	1	0
14	0	0	1	1	0	0	1	0	0	0	1	1	0	0	1	1	0	0	1	1	0	1	0	1	0	0	1	1	0	0	1	1	0	0	1	1	0	1	0	1	0	0	1	1	0	0	1	0	0	0	1	1	0	1	0	1	0	0	1	1	0	0	1	0	0	0	1	1	0	0	1	1
15	0	0	1	1	0	0	1	1	0	0	1	1	0	1	0	1	0	0	1	1	0	0	1	0	0	0	1	1	0	1	0	1	0	0	1	1	0	0	1	0	0	0	1	1	0	0	1	1	0	0	1	1	0	0	1	0	0	0	1	1	0	0	1	1	0	0	1	1	0	1	0	1
16	0	0	1	1	0	0	1	0	0	0	1	1	0	0	1	1	0	0	1	1	0	1	0	1	0	0	1	1	0	0	1	1	0	0	1	1	0	1	0	1	0	0	1	1	0	0	1	0	0	0	1	1	0	1	0	1	0	0	1	1	0	0	1	0	0	0	1	1	0	0	1	1
17	0	0	1	1	0	0	1	1	0	0	1	1	0	1	0	1	0	0	1	1	0	0	1	0	0	0	1	1	0	1	0	1	0	0	1	1	0	0	1	0	0	0	1	1	0	0	1	1	0	0	1	1	0	0	1	0	0	0	1	1	0	0	1	1	0	0	1	1	0	1	0	1
18	0	0	1	1	0	1	0	1	0	0	1	1	0	0	1	0	0	0	1	1	0	0	1	1	0	0	1	1	0	0	1	0	0	0	1	1	0	0	1	1	0	0	1	1	0	1	0	1	0	0	1	1	0	0	1	1	0	0	1	1	0	1	0	1	0	0	1	1	0	0	1	0
19	0	0	1	1	0	1	0	1	0	0	1	1	0	0	1	0	0	0	1	1	0	0	1	1	0	0	1	1	0	0	1	0	0	0	1	1	0	0	1	1	0	0	1	1	0	1	0	1	0	0	1	1	0	0	1	1	0	0	1	1	0	1	0	1	0	0	1	1	0	0	1	0
20	0	0	1	1	0	0	1	0	0	0	1	1	0	0	1	1	0	0	1	1	0	1	0	1	0	0	1	1	0	0	1	1	0	0	1	1	0	1	0	1	0	0	1	1	0	0	1	0	0	0	1	1	0	1	0	1	0	0	1	1	0	0	1	0	0	0	1	1	0	0	1	1
21	0	0	1	1	0	0	1	1	0	0	1	1	0	1	0	1	0	0	1	1	0	0	1	0	0	0	1	1	0	1	0	1	0	0	1	1	0	0	1	0	0	0	1	1	0	0	1	1	0	0	1	1	0	0	1	0	0	0	1	1	0	0	1	1	0	0	1	1	0	1	0	1
22	0	0	1	1	0	0	1	0	0	0	1	1	0	0	1	1	0	0	1	1	0	1	0	1	0	0	1	1	0	0	1	1	0	0	1	1	0	1	0	1	0	0	1	1	0	0	1	0	0	0	1	1	0	1	0	1	0	0	1	1	0	0	1	0	0	0	1	1	0	0	1	1
23	0	0	1	1	0	0	1	1	0	0	1	1	0	1	0	1	0	0	1	1	0	0	1	0	0	0	1	1	0	1	0	1	0	0	1	1	0	0	1	0	0	0	1	1	0	0	1	1	0	0	1	1	0	0	1	0	0	0	1	1	0	0	1	1	0	0	1	1	0	1	0	1
24	0	0	1	1	0	1	0	1	0	0	1	1	0	0	1	0	0	0	1	1	0	0	1	1	0	0	1	1	0	0	1	0	0	0	1	1	0	0	1	1	0	0	1	1	0	1	0	1	0	0	1	1	0	0	1	1	0	0	1	1	0	1	0	1	0	0	1	1	0	0	1	0
25	0	0	1	1	0	0	1	1	0	0	1	1	0	1	0	1	0	0	1	1	0	0	1	0	0	0	1	1	0	1	0	1	0	0	1	1	0	0	1	0	0	0	1	1	0	0	1	1	0	0	1	1	0	0	1	0	0	0	1	1	0	0	1	1	0	0	1	1	0	1	0	1
26	0	0	1	1	0	1	0	1	0	0	1	1	0	0	1	0	0	0	1	1	0	0	1	1	0	0	1	1	0	0	1	0	0	0	1	1	0	0	1	1	0	0	1	1	0	1	0	1	0	0	1	1	0	0	1	1	0	0	1	1	0	1	0	1	0	0	1	1	0	0	1	0
27	0	0	1	1	0	0	1	0	0	0	1	1	0	0	1	1	0	0	1	1	0	1	0	1	0	0	1	1	0	0	1	1	0	0	1	1	0	1	0	1	0	0	1	1	0	0	1	0	0	0	1	1	0	1	0	1	0	0	1	1	0	0	1	0	0	0	1	1	0	0	1	1

This is where the vision ended—a giant block of binary code presented to a man who knows nothing about binary code. A man goes to bed grappling with the idea of two-way ET communications and has visions that end like this. Unlike the visions of other people we discussed that gave solutions to very complex scientific problems, this vision ended with only a map to follow and no solutions at all. The journey was not complete. The fact that it got to this point was sufficient motivation for me to follow the trail and see where it led. To do this, it was necessary to analyze this block for letters. It had to be converted into letters and words.

We analyzed each row and column for letters of the alphabet. This method of analysis is different from a typical search for words and letters in a binary block. It does not read like a book. The analysis was done by looking for all the letters of the alphabet in uppercase and lowercase, as the binary values are different for each case. To get our results, we tested each column and row for every letter of the alphabet. We did not start at the top, find a letter, and then keep going. We examined the entire block for one letter, then cleaned the slate and examined the entire block for the next letter. This resulted in letter overlap. For example:

The letter *A* begins with 011 = 01100001, and the letter *C* ends with 011 = 01000011

The common/overlap digits are highlighted in yellow. When we take this to the binary block, it looks like this:

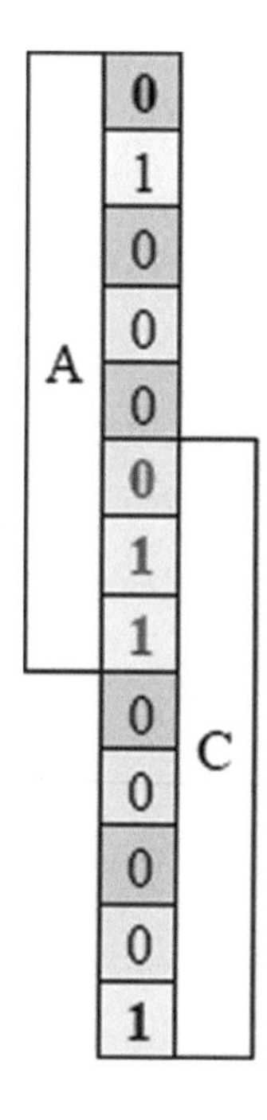

This is a very different approach from the traditional way of reading a block of code. This is deliberate, as we are trying to "think differently" as discussed earlier. Our block is a three-tiered binary block with 72 columns and 27 rows. This is a 1,944-digit block. The Arecibo block was a 1,679-digit message (265 digits smaller). Based on that, one would expect anything that came out of our block to be at least as complicated as the Arecibo block, but it is not. The result is much simpler. The important difference is the methodology used to interpret our block.

All the letters (capital and lowercase) in alphabetical order from vertical and horizontal analyses of the binary block yield fifteen letters as shown below. Note that of the letters that were found, they were found only one time. Also, recall that there were fifteen circles highlighted in our analysis of Metatron's Cube. Is this a coincidence, or is it another link to our means and methods?

A, C, d, E, F, f, i, P, Q, S, T, u, W, y, z

Now, of course, this result raises the question: how could ETs in ancient times—or at any time—put together some coded message that results in letters from the English alphabet that mean anything to us today or at any time in our history? The Bible and other ancient texts are full of references to God or gods or angels speaking in plain language to all the peoples on earth in all languages. One of many examples comes from the Bible, Acts (of the Apostles) 2:2–11:

> And suddenly there came a sound from heaven as of a rushing mighty wind, and it filled all the house where they were sitting. And there appeared unto them cloven tongues like as of fire, and it sat upon each of them. And they were all filled with the Holy Ghost, and began to speak with other tongues, as the Spirit gave them utterance. And there were dwelling at Jerusalem Jews, devout men, out of every nation under heaven. Now when this was noised abroad, the multitude came together, and were confounded, because that every man heard them speak in his own language. And they were all amazed and marveled, saying one to another, Behold, are not all these which speak Galileans? And how hear we every man in our own tongue, wherein we were born? Parthians, and Medes, and Elamites, and the dwellers in Mesopotamia, and in Judaea, and Cappadocia, in Pontus, and Asia, Phrygia, and Pamphylia, in Egypt, and in the parts of Libya about Cyrene, and strangers of Rome, Jews and proselytes, Cretes and Arabians, we do hear them speak in our tongues the wonderful works of God.

It would seem that there was no language barrier between humans and God, angels, or ETs in ancient biblical times, and there never has been. There is evidence to suggest that the ETs that came to Earth could communicate with the peoples of Earth in their own Earth language. Furthermore, there is evidence to suggest that ETs have intervened in the physical and intellectual development of humans over the entire course of human history. Perhaps that intervention included some intervention in language development.

We have followed a trail that has developed a 1,944-digit block of binary code and reduced it to 15 letters of the alphabet and still there is nothing that leaps out as any sort of method of establishing two-way communications with ETs. The analysis is still not complete. We need to analyze what these letters mean.

I put the fifteen letters into various word-scrambler programs designed to make as many words as possible out of all the possible letter combinations. After I ran all those letters through a word-scrambler program to identify only the three-letter word combinations (because prime number 3 is our main driver), 194 combinations were returned from the 15 letters. Note that different word-scramble programs can produce slight variations in the list. For this analysis, three different word-scramble programs produced the results shown in *Angel Communication Code* either exactly or within two or three placements. All the results were run through this process to get to the results that showed the most promise with respect to what is being discussed in *Angel Communication Code*. The results are as follows:

1	ACE	51	EFS	101	PYA	151	TIZ
2	ACT	52	EFT	102	PYE	152	TUI
3	ADS	53	ESQ	103	QAT	153	TUP
4	ADZ	54	ETA	104	QIS	154	TWA
5	AES	55	FAD	105	QUA	155	TYE
6	AFF	56	FAP	106	QUE	156	UPS
7	AFT	57	FAS	107	QUI	157	USE
8	AID	58	FAT	108	SAC	158	UTA
9	AIS	59	FAY	109	SAD	159	UTE
10	AIT	60	FED	110	SAE	160	UTS
11	APE	61	FET	111	SAI	161	WAD
12	APT	62	FEU	112	SAP	162	WAE
13	ASP	63	FEW	113	SAT	163	WAP
14	ATE	64	FEY	114	SAU	164	WAS
15	ATS	65	FEZ	115	SAW	165	WAT
16	AWE	66	FID	116	SAY	166	WAY
17	AYE	67	FIE	117	SAZ	167	WAZ
18	AYS	68	FIT	118	SEA	168	WEI
19	CAD	69	FIZ	119	SEC	169	WET
20	CAP	70	FUD	120	SEI	170	WEY
21	CAT	71	ICE	121	SEQ	171	WIS
22	CAW	72	ICY	122	SET	172	WIT
23	CAY	73	IDE	123	SEW	173	WIZ
24	CEP	74	IDS	124	SEZ	174	WUZ
25	CIS	75	IFF	125	SIC	175	WYE
26	CUD	76	IFS	126	SIP	176	YAP
27	CUE	77	IST	127	SIT	177	YAS
28	CUI	78	ITS	128	SPA	178	YAW
29	CUP	79	PAC	129	SPY	179	YEA
30	CUT	80	PAD	130	STY	180	YEP
31	DIA	81	PAS	131	SUE	181	YES
32	DAP	82	PAT	132	SUI	182	YET
33	DAW	83	PAW	133	SUP	183	YEW
34	DAY	84	PAY	134	SUQ	184	YEZ
35	DEF	85	PEA	135	SWY	185	YIP
36	DEW	86	PEC	136	TAD	186	YIZ
37	DEY	87	PES	137	TAE	187	YUP
38	DIE	88	PET	138	TAI	188	YUS
39	DIF	89	PEW	139	TAP	189	ZAC
40	DIP	90	PIA	140	TAS	190	ZAP
41	DIS	91	PIC	141	TAU	191	ZAS
42	DIT	92	PIE	142	TAW	192	ZES
43	DUE	93	PIS	143	TEA	193	ZIP
44	DUI	94	PIT	144	TED	194	ZIT
45	DUP	95	PIU	145	TEW		
46	DYE	96	PSI	146	TFW		
47	EAT	97	PUD	147	TIC		
48	EAU	98	PUS	148	TIE		
49	ECU	99	PUT	149	TIP		
50	EFF	100	PUY	150	TIS		

Once again, nothing immediately leaps out of the test that appears to be helpful, informative, or meaningful to our search. One hundred ninety-four combinations are many combinations to think about when attempting to decipher some sort of simple code. There was yet another step to take. Staying within the boundary conditions and the premise that the ETs' message would need to be simple and controlled by the number 3, we can drill down deeper into the list, again by using the first three prime numbers and the number 3. If we take the alphabetical list of three-letter words and acronyms and run their number positions in the list through three divisions of three, it looks like this:

3	2	1	Start	
0.03704	0.11111	0.33333	1	ACE
0.07407	0.22222	0.66667	2	ACT
0.11111	0.33333	1	3	ADS
0.14815	0.44444	1.33333	4	ADZ
0.18519	0.55556	1.66667	5	AES
0.22222	0.66667	2	6	AFF
0.25926	0.77778	2.33333	7	AFT
0.2963	0.88889	2.66667	8	AID
0.33333	1	3	9	AIS
0.37037	1.11111	3.33333	10	AIT
0.40741	1.22222	3.66667	11	APE
0.44444	1.33333	4	12	APT
0.48148	1.44444	4.33333	13	ASP
0.51852	1.55556	4.66667	14	ATE
0.55556	1.66667	5	15	ATS
0.59259	1.77778	5.33333	16	AWE
0.62963	1.88889	5.66667	17	AYE
0.66667	2	6	18	AYS
0.7037	2.11111	6.33333	19	CAD
0.74074	2.22222	6.66667	20	CAP
0.77778	2.33333	7	21	CAT
0.81481	2.44444	7.33333	22	CAW
0.85185	2.55556	7.66667	23	CAY
0.88889	2.66667	8	24	CEP
0.92593	2.77778	8.33333	25	CIS
0.96296	2.88889	8.66667	26	CUD
1	3	9	27	CUE
1.03704	3.11111	9.33333	28	CUI
1.07407	3.22222	9.66667	29	CUP
1.11111	3.33333	10	30	CUT
1.14815	3.44444	10.3333	31	DIA
1.18519	3.55556	10.6667	32	DAP
1.22222	3.66667	11	33	DAW
1.25926	3.77778	11.3333	34	DAY
1.2963	3.88889	11.6667	35	DEF
1.33333	4	12	36	DEW
1.37037	4.11111	12.3333	37	DEY
1.40741	4.22222	12.6667	38	DIE
1.44444	4.33333	13	39	DIF
1.48148	4.44444	13.3333	40	DIP
1.51852	4.55556	13.6667	41	DIS
1.55556	4.66667	14	42	DIT
1.59259	4.77778	14.3333	43	DUE
1.62963	4.88889	14.6667	44	DUI
1.66667	5	15	45	DUP
1.7037	5.11111	15.3333	46	DYE
1.74074	5.22222	15.6667	47	EAT
1.77778	5.33333	16	48	EAU
1.81481	5.44444	16.3333	49	ECU
1.85185	5.55556	16.6667	50	EFF
1.88889	5.66667	17	51	EFS
1.92593	5.77778	17.3333	52	EFT
1.96296	5.88889	17.6667	53	ESQ
2	6	18	54	ETA
2.03704	6.11111	18.3333	55	FAD
2.07407	6.22222	18.6667	56	FAP
2.11111	6.33333	19	57	FAS
2.14815	6.44444	19.3333	58	FAT
2.18519	6.55556	19.6667	59	FAY
2.22222	6.66667	20	60	FED
2.25926	6.77778	20.3333	61	FET
2.2963	6.88889	20.6667	62	FEU
2.33333	7	21	63	FEW
2.37037	7.11111	21.3333	64	FEY
2.40741	7.22222	21.6667	65	FEZ
2.44444	7.33333	22	66	FID
2.48148	7.44444	22.3333	67	FIE
2.51852	7.55556	22.6667	68	FIT
2.55556	7.66667	23	69	FIZ
2.59259	7.77778	23.3333	70	FUD
2.62963	7.88889	23.6667	71	ICE
2.66667	8	24	72	ICY
2.7037	8.11111	24.3333	73	IDE
2.74074	8.22222	24.6667	74	IDS
2.77778	8.33333	25	75	IFF
2.81481	8.44444	25.3333	76	IFS
2.85185	8.55556	25.6667	77	IST
2.88889	8.66667	26	78	ITS
2.92593	8.77778	26.3333	79	PAC
2.96296	8.88889	26.6667	80	PAD
3	9	27	81	PAS
3.03704	9.11111	27.3333	82	PAT
3.07407	9.22222	27.6667	83	PAW
3.11111	9.33333	28	84	PAY
3.14815	9.44444	28.3333	85	PEA
3.18519	9.55556	28.6667	86	PEC
3.22222	9.66667	29	87	PES
3.25926	9.77778	29.3333	88	PET
3.2963	9.88889	29.6667	89	PEW
3.33333	10	30	90	PIA
3.37037	10.1111	30.3333	91	PIC
3.40741	10.2222	30.6667	92	PIE
3.44444	10.3333	31	93	PIS
3.48148	10.4444	31.3333	94	PIT
3.51852	10.5556	31.6667	95	PIU
3.55556	10.6667	32	96	PSI
3.59259	10.7778	32.3333	97	PUD
3.62963	10.8889	32.6667	98	PUS
3.66667	11	33	99	PUT
3.7037	11.1111	33.3333	100	PUY

3.74074	11.2222	33.6667	101	PYA
3.77778	11.3333	34	102	PYE
3.81481	11.4444	34.3333	103	QAT
3.85185	11.5556	34.6667	104	QIS
3.88889	11.6667	35	105	QUA
3.92593	11.7778	35.3333	106	QUE
3.96296	11.8889	35.6667	107	QUI
4	12	36	108	SAC
4.03704	12.1111	36.3333	109	SAD
4.07407	12.2222	36.6667	110	SAE
4.11111	12.3333	37	111	SAI
4.14815	12.4444	37.3333	112	SAP
4.18519	12.5556	37.6667	113	SAT
4.22222	12.6667	38	114	SAU
4.25926	12.7778	38.3333	115	SAW
4.2963	12.8889	38.6667	116	SAY
4.33333	13	39	117	SAZ
4.37037	13.1111	39.3333	118	SEA
4.40741	13.2222	39.6667	119	SEC
4.44444	13.3333	40	120	SEI
4.48148	13.4444	40.3333	121	SEQ
4.51852	13.5556	40.6667	122	SET
4.55556	13.6667	41	123	SEW
4.59259	13.7778	41.3333	124	SEZ
4.62963	13.8889	41.6667	125	SIC
4.66667	14	42	126	SIP
4.7037	14.1111	42.3333	127	SIT
4.74074	14.2222	42.6667	128	SPA
4.77778	14.3333	43	129	SPY
4.81481	14.4444	43.3333	130	STY
4.85185	14.5556	43.6667	131	SUE
4.88889	14.6667	44	132	SUI
4.92593	14.7778	44.3333	133	SUP
4.96296	14.8889	44.6667	134	SUQ
5	15	45	135	SWY
5.03704	15.1111	45.3333	136	TAD
5.07407	15.2222	45.6667	137	TAE
5.11111	15.3333	46	138	TAI
5.14815	15.4444	46.3333	139	TAP
5.18519	15.5556	46.6667	140	TAS
5.22222	15.6667	47	141	TAU
5.25926	15.7778	47.3333	142	TAW
5.2963	15.8889	47.6667	143	TEA
5.33333	16	48	144	TED
5.37037	16.1111	48.3333	145	TEW
5.40741	16.2222	48.6667	146	TFW
5.44444	16.3333	49	147	TIC
5.48148	16.4444	49.3333	148	TIE
5.51852	16.5556	49.6667	149	TIP
5.55556	16.6667	50	150	TIS

5.59259	16.7778	50.3333	151	TIZ
5.62963	16.8889	50.6667	152	TUI
5.66667	17	51	153	TUP
5.7037	17.1111	51.3333	154	TWA
5.74074	17.2222	51.6667	155	TYE
5.77778	17.3333	52	156	UPS
5.81481	17.4444	52.3333	157	USE
5.85185	17.5556	52.6667	158	UTA
5.88889	17.6667	53	159	UTE
5.92593	17.7778	53.3333	160	UTS
5.96296	17.8889	53.6667	161	WAD
6	18	54	162	WAE
6.03704	18.1111	54.3333	163	WAP
6.07407	18.2222	54.6667	164	WAS
6.11111	18.3333	55	165	WAT
6.14815	18.4444	55.3333	166	WAY
6.18519	18.5556	55.6667	167	WAZ
6.22222	18.6667	56	168	WEI
6.25926	18.7778	56.3333	169	WET
6.2963	18.8889	56.6667	170	WEY
6.33333	19	57	171	WIS
6.37037	19.1111	57.3333	172	WIT
6.40741	19.2222	57.6667	173	WIZ
6.44444	19.3333	58	174	WUZ
6.48148	19.4444	58.3333	175	WYE
6.51852	19.5556	58.6667	176	YAP
6.55556	19.6667	59	177	YAS
6.59259	19.7778	59.3333	178	YAW
6.62963	19.8889	59.6667	179	YEA
6.66667	20	60	180	YEP
6.7037	20.1111	60.3333	181	YES
6.74074	20.2222	60.6667	182	YET
6.77778	20.3333	61	183	YEW
6.81481	20.4444	61.3333	184	YEZ
6.85185	20.5556	61.6667	185	YIP
6.88889	20.6667	62	186	YIZ
6.92593	20.7778	62.3333	187	YUP
6.96296	20.8889	62.6667	188	YUS
7	21	63	189	ZAC
7.03704	21.1111	63.3333	190	ZAP
7.07407	21.2222	63.6667	191	ZAS
7.11111	21.3333	64	192	ZES
7.14815	21.4444	64.3333	193	ZIP
7.18519	21.5556	64.6667	194	ZIT

The result after the third round of division by 3 (column 3, on the left) is that there are four words (highlighted in green) that are divisible by 3 down to the first three prime numbers—2, 3, and 5—plus the fourth prime number, which is 7. Now we are getting somewhere. The word search filtering and prioritizing results are as follows:

2 = ETA; 3 = PAS; 5 = SWY; 7 = ZAC

Let us investigate these words in the order in which they dropped out of the binary block analysis. The number 7 has significance in that 7 is the next prime number after the 2–3–5 binary block analysis that got us to this point. We put the 2–3–5 number sequence into the machine and turned the crank, and we got back words that correspond to 2–3–5 plus 7. Suddenly the appearance of the number 7 from our deciphering efforts links up to all the references to the number 7 being of historical significance previously discussed. This is an important connection.

The seventh letter of the Greek alphabet is eta. It originally meant the number 7 also. Eta is also the first three-letter group turned out by the model that corresponds above to the first prime number (2). The appearance of the number 7 also seems to close the loop, corresponding to the word *Zac*. *Eta* is the first word to drop out of the vision analysis, and eta the number is the last relevant output from our block. It is the beginning and the end, the alpha and the omega of our analysis. In addition, the Bible tells us: "So the last shall be first, and the first last: for many be called, but few chosen" (Matthew 20:16).

God "chose" people to do certain things several times in the Bible, and He delivered the instructions for what they were to do either directly or through angels. Many times the message was received in a dream or vision. A dream vision is exactly how we got to this point in *Angel Communication Code* and how many people throughout history made very important discoveries as previously discussed.

In our code, the number 7 correlates to the three-letter word *Zac*. Zac is an ancient Hebrew boy's name that means "the Lord has remembered." Is the number 7 that shows up in our process at this point a clue that the code has ended on this last three-letter item? The fact that the number 7 has dropped out of our process is significant.

The clues being revealed are screaming out loud that we are definitely on the right trail. Something is going on here. With the completion of each step in the process, we are given more clues that suggest we should keep going.

Let us look closer at these words, stating with the first word, *eta*.

Greek Alphabet Eta

As noted previously, eta is the seventh letter of the Greek alphabet. It was also originally the Greek number 7. The seven brightest stars in the Orion constellation form an hourglass shape.

An hourglass is an ancient tool from the Middle Ages used repeatedly to measure the passage of a certain amount of time. The constellation Orion, and more specifically, Orion's belt, is known to have significance in many ancient cultures and is believed to have been used for the construction and positioning of their megalithic structures. We have already discussed the importance of the number 7 in our process so far. Much of what we know about geometry and mathematics comes from the ancient Greek mathematicians. The word *mathematics* itself is derived from the Greek *mathema*, meaning "the subject of instruction." We are using these principles (instructions) to guide us through a process to find our way to the ETs.

International Morse Code Eta

Much has been already been written about Samuel Morse, the inventor and composer of international Morse code. His entire legacy will not be rewritten here, but know that he was a fascinating man. The simplistic overview as it applies to our subject is that Morse was the first person to standardize a code for communications originally sent down a wire that could be decoded at the other end. It is also transmitted visually with lights or wirelessly via radio. Morse knew he wanted to invent a code that could translate electrical signals into letters and then words. That code had to be simple to understand and use. To simplify his code, he set out to determine the most frequently used letters of the alphabet. He did not go to the dictionary to count letters; rather, he counted the most frequently used letters in printer's type for common publications and news bulletins, etc. This made sense because most of the words we use are simple and are repeated frequently in conversation and in printed communications.

Morse's search results were as follows, in order of frequency of use:

#1 = E #2 = T #3 = A

Morse code has nothing to do with the derivation of the code that is the subject of *Angel Communication Code*; however, eta is of great significant to both. This is probably not a coincidence.

Morse code is a code of communication that is internationally recognized by all developed countries that have communication capabilities.[8]

Let us take a closer look at this and note some relevant observations. International Morse code looks like this:

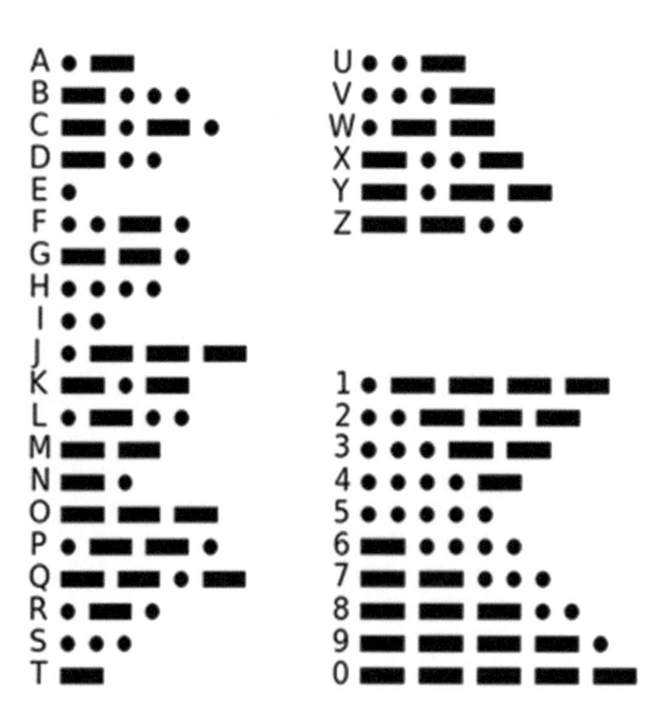

8 *New World Encyclopedia*, s.v. "Morse Code," https://www.newworldencyclopedia.org/entry/Morse_Code, accessed November 24, 2023.

1. There are only two letters that have single symbols—the *e* (one dot) and the *t* (one dash).
2. If you put the letters *e* and *t* together, you get *a* (dot dash).
3. If you put the letters *e*, *t*, and *a* together, you get dot dash dot dash. This is the simplest alternating pattern possible.
4. In keeping with our focus on the number 3, notice that there are only two letters that use a symbol three times alone with no other symbol—the *s* (dot dot dot) and the *o* (dash dash dash). Put these letters together in a repeating pattern and you get SOS, the international distress signal. It means either "save our ship" or "save our souls."

Another relevant point having to do with the mechanics of how a standard message is transmitted in Morse code is that a dot is considered one "unit" and a dash is considered three "units." This means that the SOS distress signal consists of fifteen units (3 + 9 + 3 = 15), not counting the spaces between. This is a connection with the fifteen Metatron's Cube circles and the fifteen-letter output from our code analysis.

When transmitting a message, the standard convention is as follows:

- The space between letters is three units.
- The space between words is seven units.

The layout of a single SOS message would therefore be made up of twenty-one units as shown below:

S			*Space*			O									*Space*			S		
1	1	1	3 Units			3			3			3			3 Units			1	1	1
●	●	●				–			–			–						●	●	●
1	2	3	4	5	6	7	8	9	10	11	12	13	14	15	16	17	18	19	20	21

If we sent out the SOS message three times and included the seven units between transmissions, it would be this:

21 units × 3 = 63 units + 7 units + 7 units = 77 total units

Now recall all the many things of significance we previously discussed about the number 7 with respect to its being part of a coded message, including Luke's description of the seventy-seven generations from God to Jesus. This is an incredible connection.

The telegraph, the first form of telecommunication in the history of humankind, was operated using the language of international Morse code. Morse code is clearly linked to what we are doing in *Angel Communication Code*. It is important to recognize that we were guided to eta via a binary block of threes and not by any counting of letter-usage frequency in written or spoken language. The discovery of our eta and Morse's eta are not linked in that way. Our eta and Morse's eta were independently derived via different methodologies, but they have the same general intent, namely, to simplify communications over long distances. That cannot be a coincidence and could be considered "independent verification."

Eta-Earth

There is a concept out there in the astrophysics world called "eta-Earth." The *Encyclopedia of Astrobiology* defines eta-Earth as follows:

> The term *eta-Earth* is defined as the mean number per star of rocky planets with between 1 and 1.5–2 Earth-radii that reside in the optimistic habitable zone (HZ) of their host star. Eta-Earth enters one formulation of the Drake equation, which endeavors to estimate the occurrence of intelligent life in the galaxy; at the present time, it is usually calculated separately for each stellar spectral type. Thus, eta-Earth represents the occurrence rate of rocky planets in the optimistic HZ of different stars. The references present some values for eta-Earth based on different statistical analyses of the data from the Kepler space telescope.

There is a lot more to it than is discussed here. The point is that eta is a clue that relates to trying to figure out the potential locations of ETs. Someone along the way called these potentially habitable planets "eta-Earths." NASA estimates that there are about one billion "earths" in our galaxy alone, so the odds that there is at least one other habitable planet out there supporting intelligent life similar to our own are strong.

The Eta Function

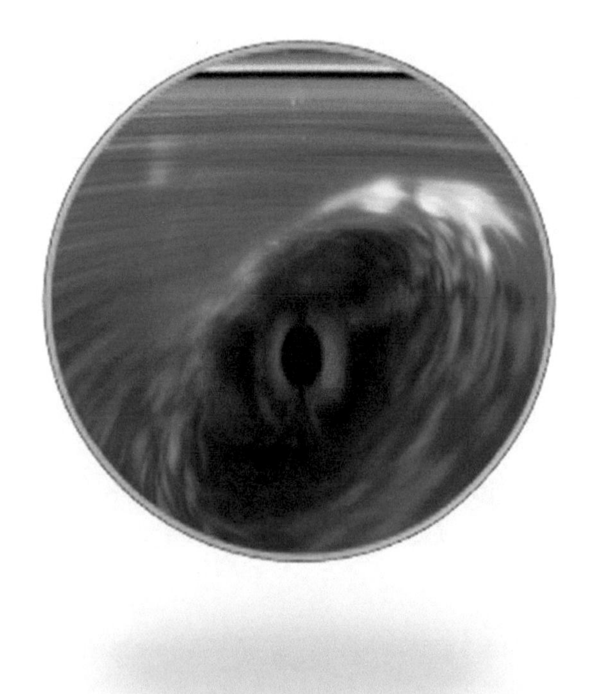

In mathematics and astrophysics, there is a concept called the eta function. It is linked to the study of string theory and the existence of wormholes.

A wormhole (also called an Einstein–Rosen bridge) is a theoretical structure in the universe that connects different points in space-time. It is based on a special solution of Einstein field equations. A wormhole can be described as a tunnel with two ends at separate locations in space and/or time. The theory relies on Einstein's general theory of relativity, and the existence of wormholes has not been proven. In theory, wormholes can connect extremely long distances (as much as a billion light-years or more), short distances (such as a few feet), different galaxies, or different points in time.[9]

Two of the questions to which we seek answers are: how did ETs get here, and where did/do they go? You can't have that conversation without a discussion about wormholes. The eta function is a tool used by some very smart people in search of an explanation, proof, or understanding of wormholes. It is now brought before us in this quest for answers via the word *eta* that fell out of our binary block analysis. That link is significant; it surely means something.

---

9 Nola Taylor Tillman and Ailsa Harvey, "What Is Wormhole Theory?," Space.com, January 13, 2022.

Eta Carinae

Eta Carinae is a much-studied gigantic stellar system that has two stars in relatively close proximity. Its mass is estimated to be 100–150 times the mass of our sun, and its luminosity is estimated to be about four million times that of our sun.

Eta Carinae lies approximately seventy-five hundred light-years from Earth, within the constellation of Carina. *Carina* is Latin for the keel of a ship. Carina used to be part of a much larger constellation that is called Argo Navis. Argo Navis, whose name is Latin for "the ship," includes the star called Puppis (meaning "stern") and the star called Vela (meaning "sails").

Eta Carinae lies within the Carina Nebula and is currently the largest star cluster that can be effectively studied in detail. This is because of its location and size. A nebula is a huge cloud of dust and gas out in space. Some nebulas are formed when a dying star explodes, and conversely, some are formed when new stars are developing.[10]

A commonly used NASA telescope image of Eta Carinae is shown below.

Eta Carinae cannot be seen north of latitude 30° north and never sets below the horizon south of latitude 30° south. It is best seen during the month of March, the month of the spring equinox. Later, you will see why this is important.

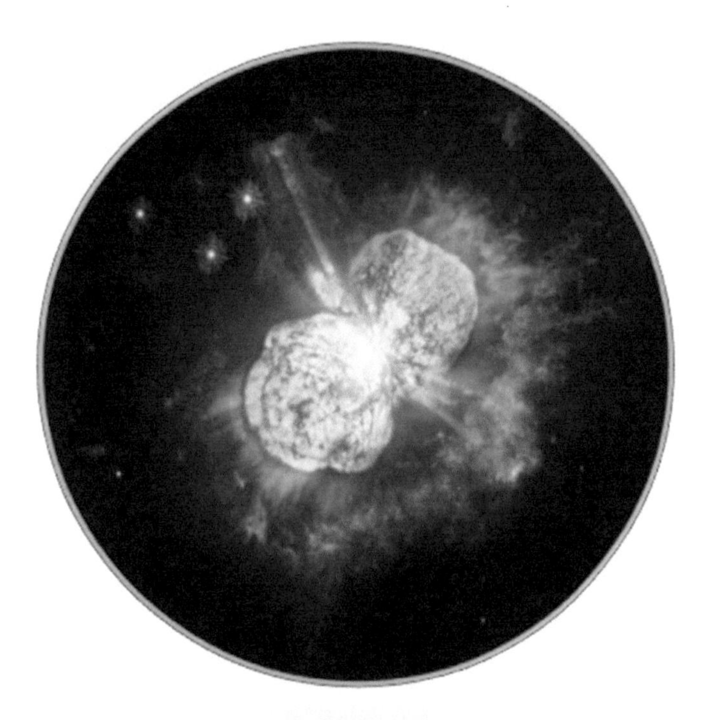

[10] Kate Davis, "Eta Carinae, Sovereign of the South," American Association of Variable Star Observers, https://www.aavso.org/vsots_etacar, accessed November 24, 2023.

Other Eta Meanings of Relevance

In human anatomy, the carina of the trachea is a ridge at the base of the windpipe separating the right and left bronchial tubes. It triggers the cough reflex and has an influence on a person's ability to speak (communicate).

The word *Eta* in Sanskrit is a girl's name that means "luminous." Sanskrit, which means "perfected" or "refined," is believed to be one of the oldest (if not *the* oldest) human languages, dating back to the second millennium BC. The liturgical language of Hinduism and Buddhism, Sanskrit was considered to be a sophisticated way of speaking and an indicator of high status and education.

Eta Islands

There is a lot of relevant information contained within the word and number *eta*. One critical and relevant fact about eta pertains to geographic locations on earth.

There are three (and only three) islands on earth called Eta Island:

1. Eta Island Canada
2. Eta Island Antarctica
3. Eta Island Bermuda.

When they are looked at independently, nothing conspicuously special about these places stands out with respect to being centers of ancient culture or having megalithic structures pointing to celestial bodies, at least as far as we know. Perhaps we should mount some expeditions to these places and look more closely. Eta Island in Canada and Eta Island in Antarctica are very remote places, and access is far from easy.

If we look at these locations together, we can easily identify some very meaningful correlations to the path on which we are traveling in *Angel Communication Code*. If you draw a line connecting these islands on a map, you find that it looks like this (outside looking in):

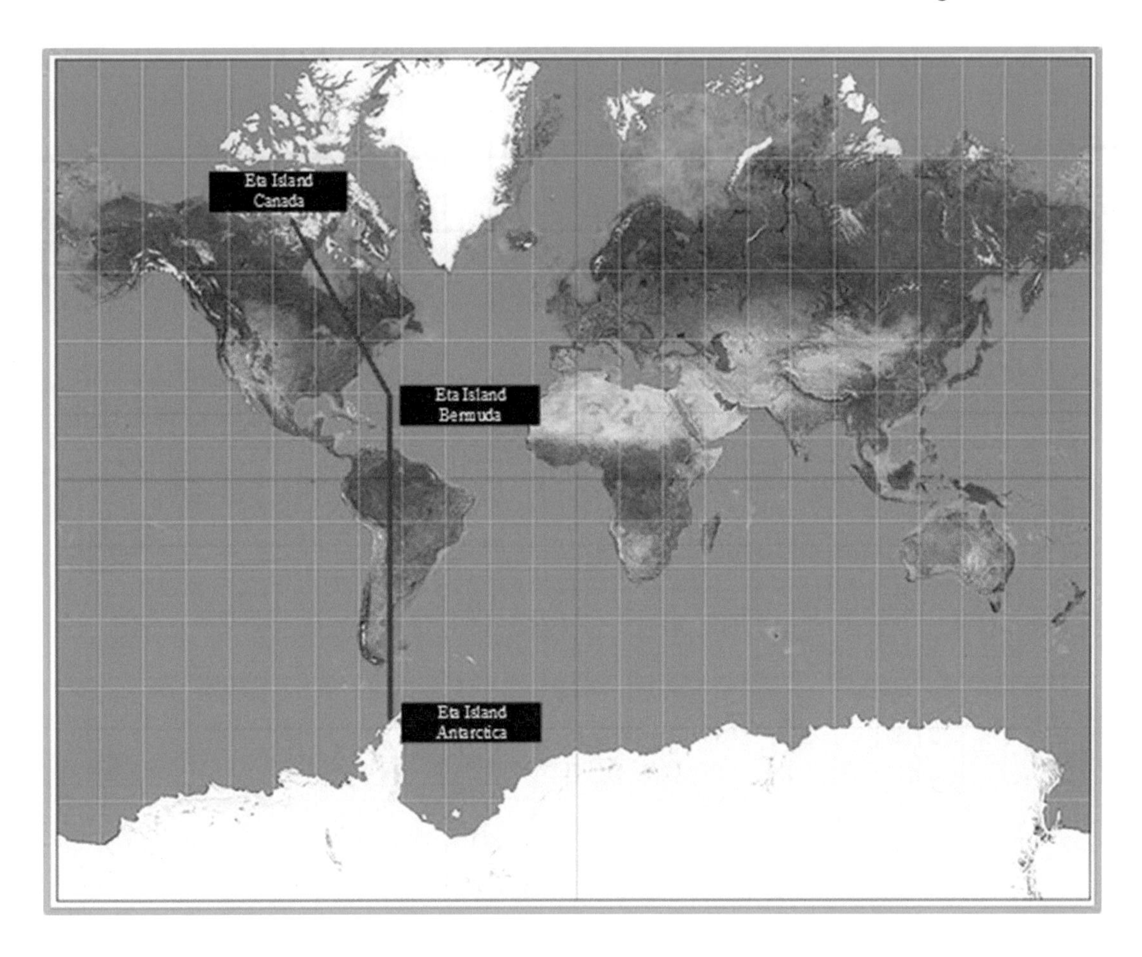

Eta Bermuda and Eta Antarctica are within about two degrees of being on the same line of longitude. One of the Earth phenomena related to the number 3 is the three stars of Orion's belt and their alignment with several ancient megalithic structures, the most famous of which are the pyramids of Egypt, shown below. This is the "inside looking out" view of the alignments:

From this angle and distance, the belt looks to be a fairly straight line. The line we drew connecting the Eta Islands is not a straight line from that angle and distance. If we draw those lines onto a more three-dimensional model, such as a Google Earth image, it looks like this:

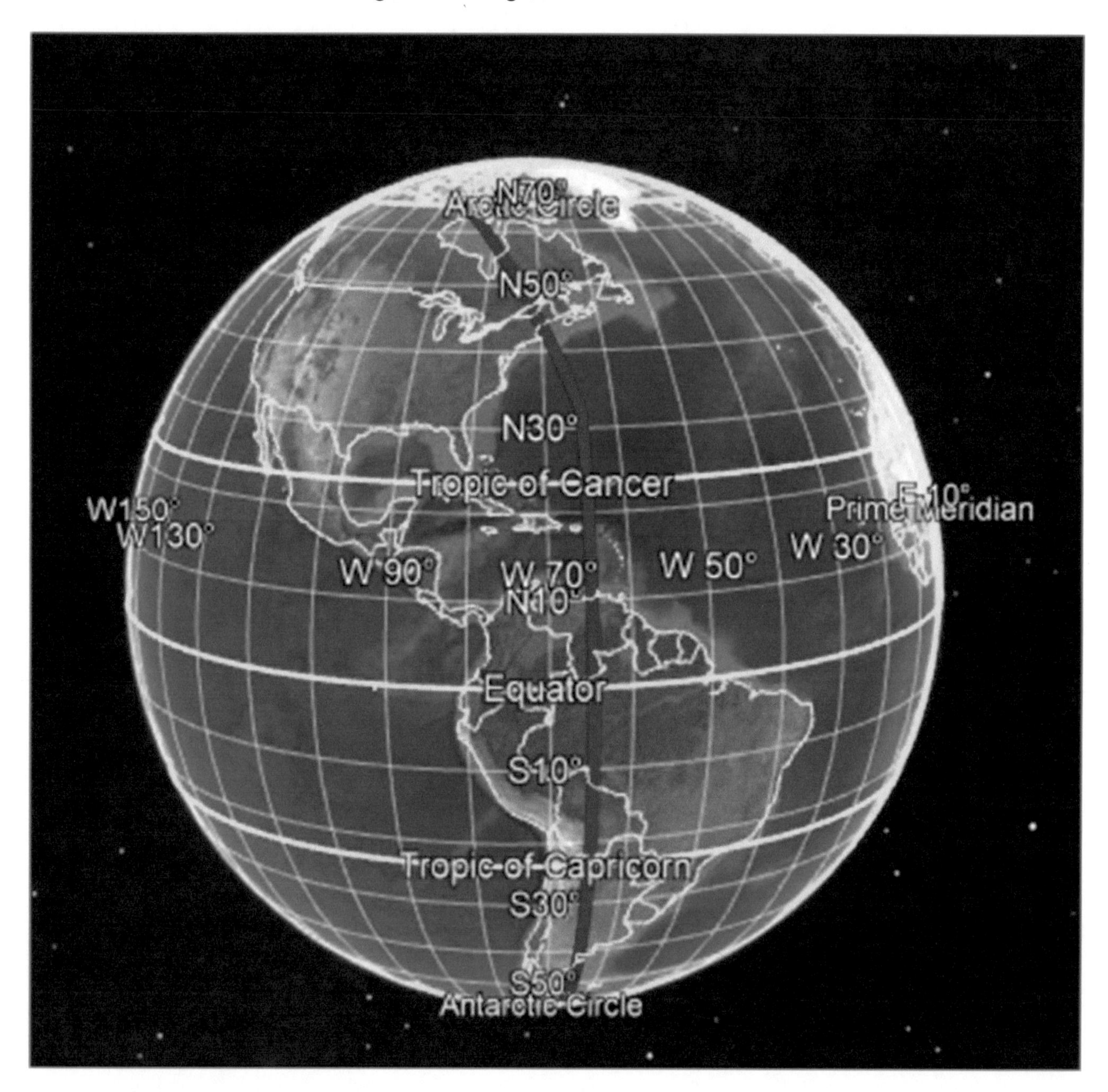

Laying it out on a sphere makes us realize that Orion's belt only looks like a straight line, or close to it, because of the viewing angle and because it's light-years away.

Following is a snapshot taken from a video of Orion shown on the Hubble website. The video, which goes on to show Orion as the camera passes around it, can be found online at https://hubblesite.org/video/5-the-true-shape-of-orion.

From this angle and distance, it looks that the stars of Orion's belt are aligned over the pyramids. This is how we always see Orion in the night sky—a fairly straight line. Notice the hourglass shape previously discussed.

Forty-six seconds into the Hubble video, however, the alignment looks like this:

Then fifty-three seconds into the video we see this:

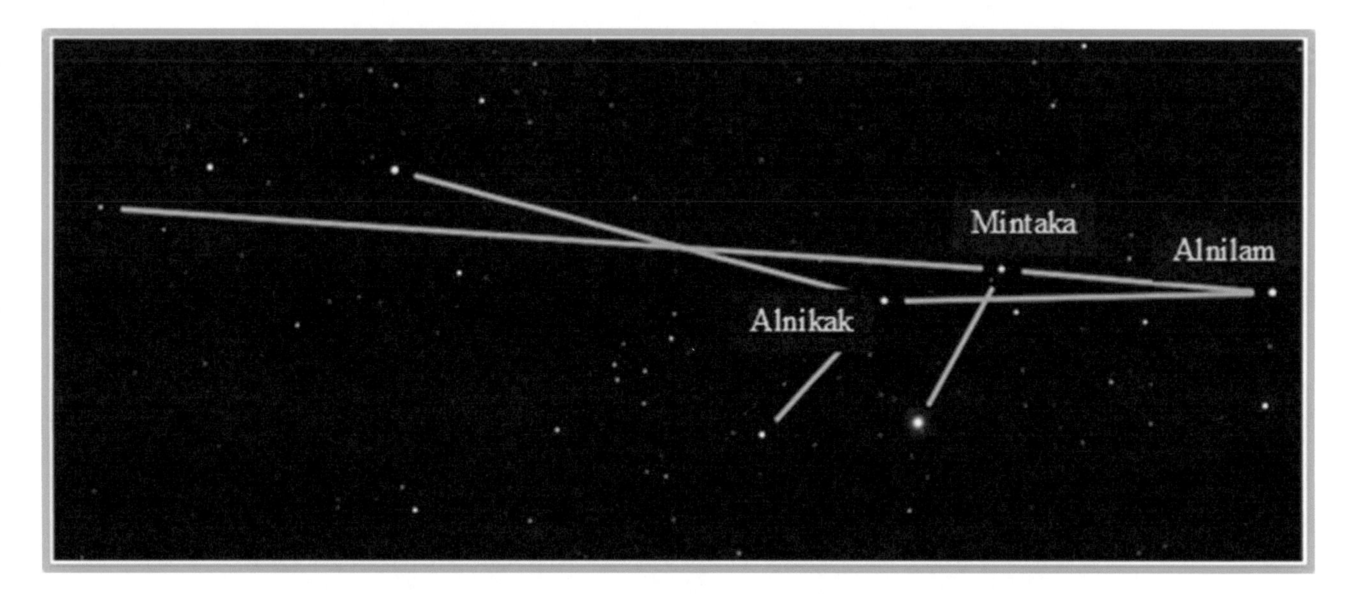

Rotate the fifty-three-second snapshot ninety degrees clockwise, and it looks a lot like our Eta Island line.

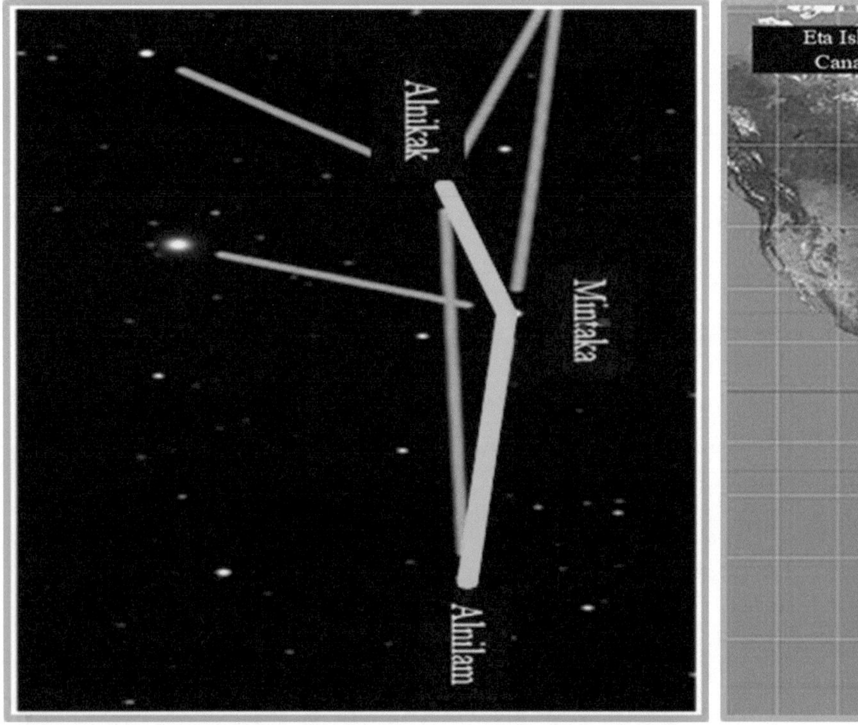

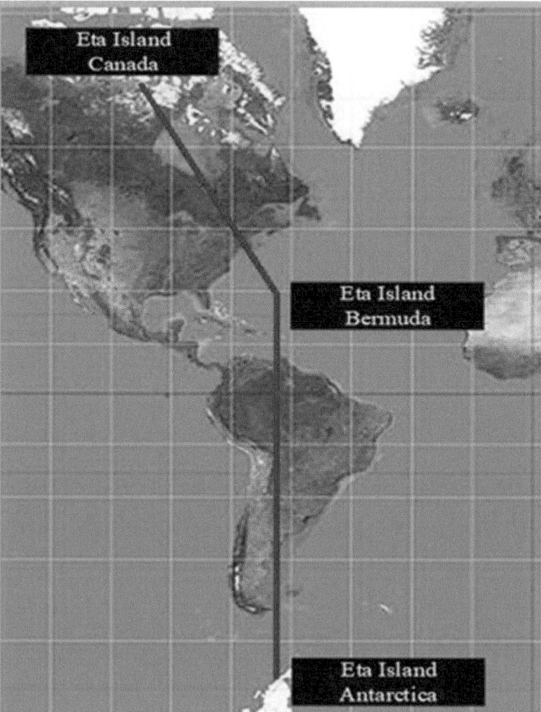

Now compare these alignments to a (Google Earth) view from directly above the pyramids in a position perpendicular to the earth.

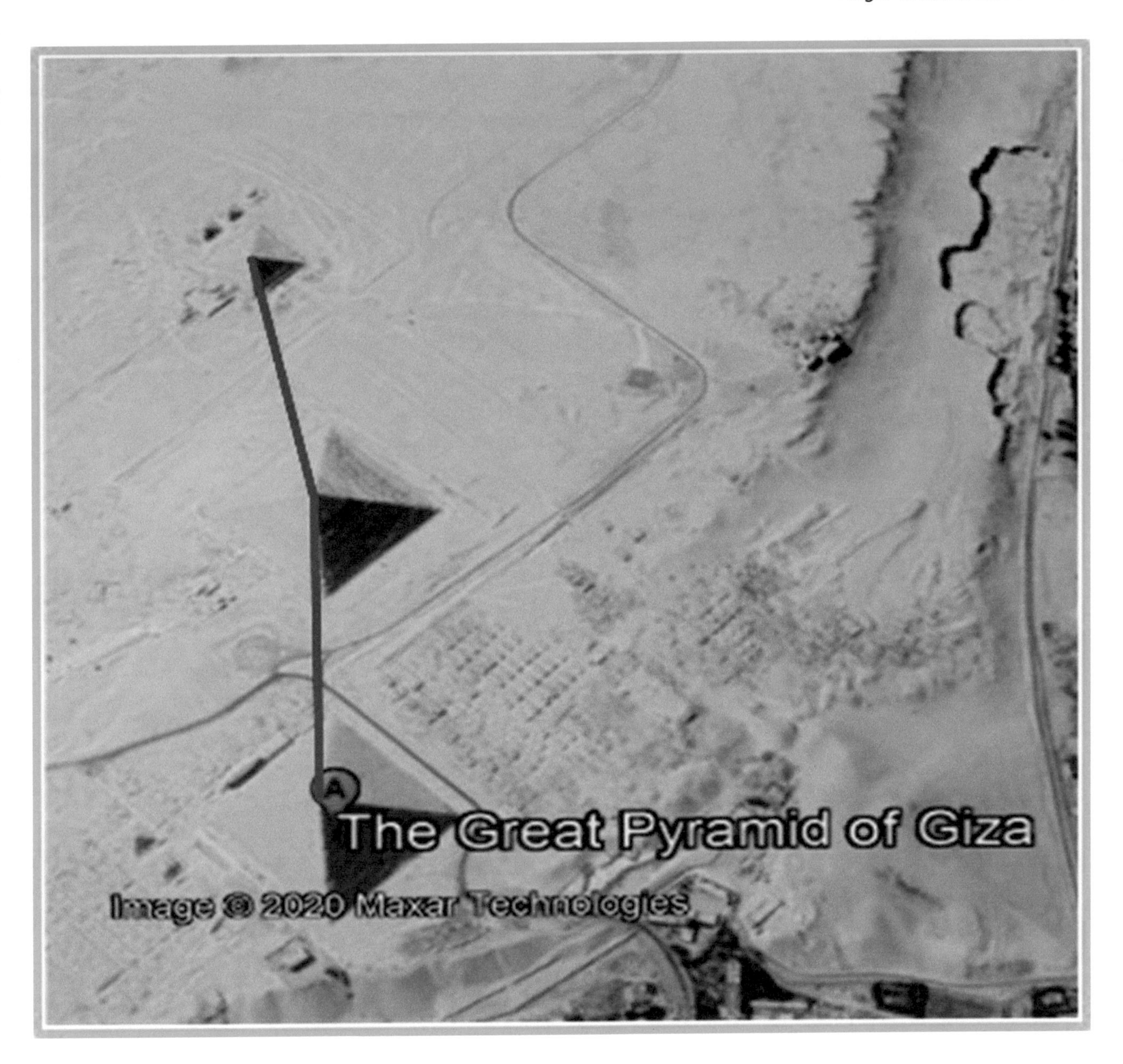

This is the outside-looking-in perspective. This relationship of alignments with Orion's belt is also found with other ancient megalithic structures from different cultures on the other side of the world. There is an indisputable relationship between the alignment of Orion's belt and that of Eta Island. What makes this finding unique is that the three Eta islands are not man-made or ET-constructed structures. They are geological features that are now revealed through my process that started with a dreamlike vision.

The point of showing the snapshots from the Hubble video, the Google Earth image alignment, and the Eta Island alignment together is to demonstrate that there are locations somewhere in the universe where ETs would see the belt of Orion as being in exactly the same alignment as the three Eta Islands. There are many points in space where this alignment would occur as the earth rotates on its axis and orbits the sun. Perhaps those outside-looking-in locations, and not Orion itself, should be the locations we target in our search. This concept has never been presented or tested to the best of this author's knowledge.

Is it possible that when Orion's belt and the three aligned Eta Islands are in some sort of synchronized alignment together, something happens that opens a gateway for interstellar travel to and from Earth?

Since the distances are measured in light-years, there has to be some sort of wormhole or similar portal that allows this to happen. The gap in space and time must somehow be closed in order to make the journey possible, and at this point, we have every reason to believe that this journey has already been be made by ETs many times. The number 7 and the word *eta* dropping out of our analysis is a very important thing.

NASA published an article in 2012 describing the results of research on "portals" that they funded at the University of Iowa. A portion of that article is shown below:

> A favorite theme of science fiction is "the portal"—an extraordinary opening in space or time that connects travelers to distant realms. A good portal is a shortcut, a guide, a door into the unknown. If only they actually existed …
>
> It turns out that they do, sort of, and a NASA-funded researcher at the University of Iowa has figured out how to find them.
>
> "We call them X-points or electron diffusion regions," explains plasma physicist Jack Scudder of the University of Iowa. "They're places where the magnetic field of Earth connects to the magnetic field of the Sun, creating an uninterrupted path leading from our own planet to the sun's atmosphere ninety-three million miles away."
>
> Observations by NASA's *THEMIS* spacecraft and Europe's *Cluster* probes suggest that these magnetic portals open and close dozens of times each day. They're typically located a few tens of thousands of kilometers from Earth where the geomagnetic field meets the onrushing solar wind. Most portals are small and short-lived; others are yawning, vast, and sustained. Tons of energetic particles can flow through the openings, heating Earth's upper atmosphere, sparking geomagnetic storms, and igniting bright polar auroras.
>
> NASA is planning a mission called "MMS," short for Magnetospheric Multiscale Mission, due to launch in 2014, to study the phenomenon. Bristling with energetic particle detectors and magnetic sensors, the four spacecraft of MMS will spread out in Earth's magnetosphere and surround the portals to observe how they work.

To the best of our knowledge, the MMS mission made many interesting discoveries that support the idea of a magnetic influence; however, it did not specifically discover any sort of portal.

What we have uniquely discovered here in *Angel Communication Code* is the following:

a. ancient megalithic structures that have this Orion alignment are exactly that: structures that were built by whomever that are deliberately aligned to the celestial bodies, and
b. our discovery is not constructed. Eta islands are naturally occurring geological features given the name Eta that have the celestial alignment. We were directed to three specific and unique points on the earth through a process based on the binary clues provided. There is a set of directions with clear meaning as to what we are being guided to find. The three Eta Islands alignment is a natural alignment, as far as we know. It is not an alignment that was constructed like the pyramids. That fact is unique and significant.

It seems highly unlikely that the links we are seeing thus far are arbitrary, coincidental, or without intelligent design. The significance of Orion's belt is clear and documented by several early and very different cultures, and that fact has to mean something with respect to what we are doing here in *Angel Communication Code*. There is now this geographic link via the Eta Islands to include in the library of evidence, and it gets more relevant as we continue to look deeper.

Within a very small margin of error, the Eta line endpoints lie on two earth circles. Now we begin to get to the geometric keys to our code: the circle and the triangle. The Canada Eta Island end of the Eta line is only a couple of degrees north of the Arctic Circle, and the Antarctic Eta Island end of the Eta line is only a couple of degrees north of the Antarctic Circle.

As a frame of reference, understand that one degree of earth latitude is equal to approximately sixty-nine miles. On a global scale, and even more so on a universal scale, two earth degrees is a miniscule distance. For all intents and purposes, the endpoints of the Eta line are riding on the Arctic and Antarctic Circles. What is so special about these earth circles?

The Arctic Circle delineates the southernmost latitude of the Northern Hemisphere, where the sun does not rise above the horizon during the day of the winter solstice. Conversely, it is the southernmost latitude of the Northern Hemisphere, where the sun does not set during the summer solstice. During these two days of the year, the sun is either just blow or just above the horizon, respectively. The opposite solar relationships occur during this same time at the Antarctic Circle latitude.

A simplistic diagram of this dynamic is shown below.

Source: Timeanddate.com.

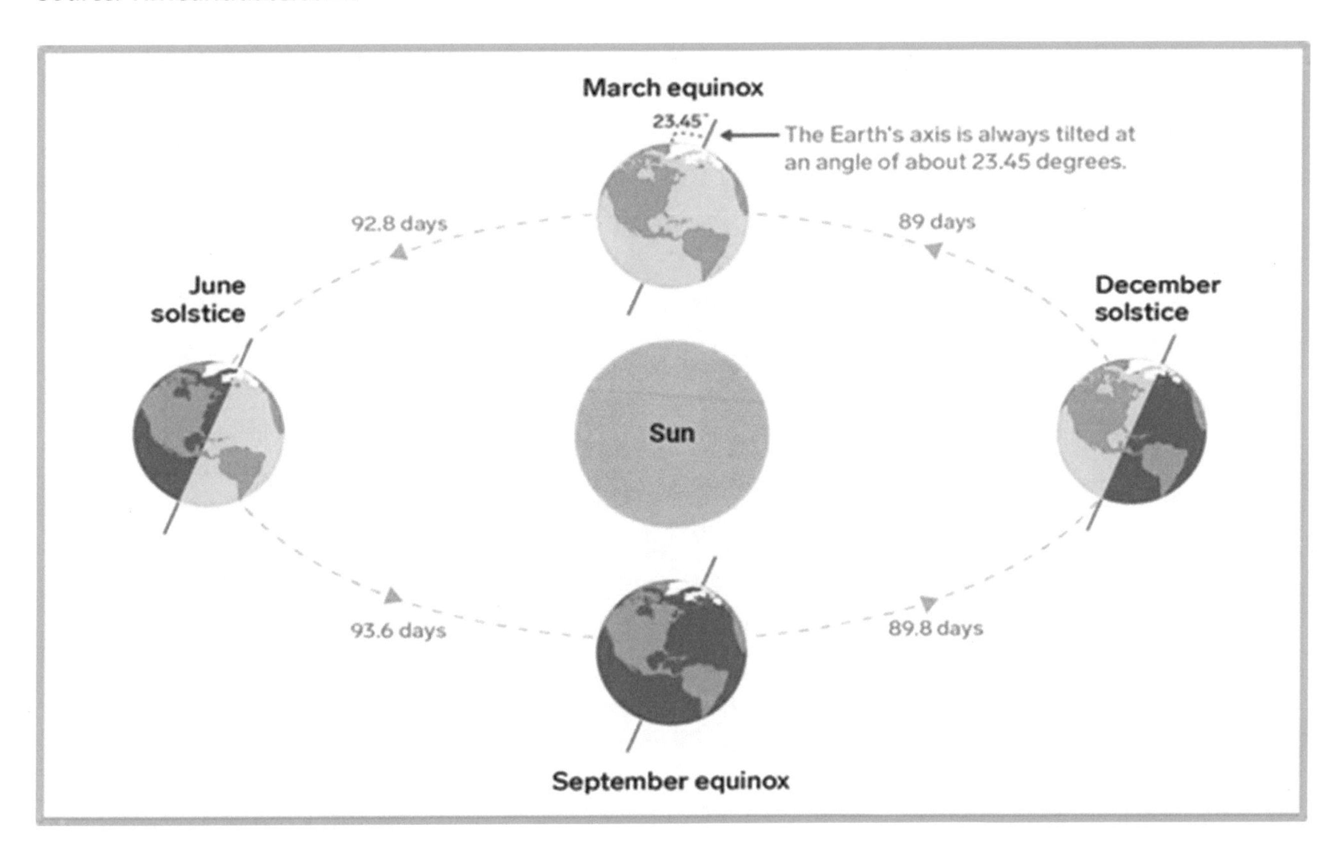

Notice that the earth is tilted on its axis by 23.45° or approximately 23.5°. The relevance of the first three prime numbers (2, 3, and 5) surfaces again.

What that means in the context of the significance of the Arctic and Antarctic Circles being the start points and endpoints of the Eta Island lines is this: they are the only two latitudes on earth where there is exactly one twenty-four-hour period of full daylight and exactly one twenty-four-hour period of twilight. They are unique delineators of night and day—one day of complete light and one day of complete darkness, perhaps an annual reminder of the alpha and the omega, the beginning and the end of our Eta lines. Recall that Eta was also the first and last (alpha and omega) output from our code analysis. This is a significant relationship.

The relationship of earth circles, equinox events, and solstice events has been critical to the development of celestial navigation equations and tables for hundreds of years. It is complicated, and I am not an astronomer, so I offer the following explanation taken from *Bowditch's American Practical Navigator* (the bible of celestial navigation) at https://www.starpath.com/celnavbook/BCN_all.pdf.

Pay close attention to the occurrences of 2, 3, 5, and 7 and to the history:

> Since the earth travels faster when nearest the sun, the Northern Hemisphere (astronomical) winter is shorter than its summer by about seven days [7 days]. Everywhere between the parallels of about 23°26′ N [23.5] and about 23°26′ S [23.5] the sun is directly overhead at some time during the year. Except at the extremes, this occurs twice: once as the sun appears to move northward and the second time as it moves southward. This is the torrid zone. The northern limit is the Tropic of Cancer, and the southern limit's the Tropic of Capricorn. These names come from the constellations which the sun entered at the solstices when the names were first applied more than two thousand years ago. Today, the sun is in the next constellation toward the west because of precession of the equinoxes. The parallels about 23°26′ [23.5] from the poles, marking the approximate limits of the circumpolar sun, are called polar circles, the one in the Northern Hemisphere being the Arctic Circle and the one in the Southern Hemisphere the Antarctic Circle. The areas inside the polar circles are the north and south frigid zones. The regions between the frigid zones and the torrid zones are the north and south temperate zones. The expression "vernal equinox" and associated expressions are applied both to the times and points of occurrence of the various phenomena. Navigationally, the vernal equinox is sometimes called the first point of Aries (symbol) because, when the name was given, the sun entered the constellation Aries, the ram, at this time. This point is of interest to navigators because it is the origin for measuring sidereal hour angle. The expressions March equinox, June solstice, September equinox, and December solstice are occasionally applied as appropriate, because the more common names are associated with the seasons in the Northern Hemisphere and are six months out of step for the Southern Hemisphere. The axis of the earth is undergoing a precessional motion similar to that of a top spinning with its axis tilted. In about 25,800 [25.7 × 1,000+/–] years the axis completes a cycle and returns to the position from which it started. Since the celestial equator is 90° from the celestial poles, it too is moving. The result is a slow westward movement of the equinoxes and solstices, which has already carried them about 30°, or one constellation, along the ecliptic from the positions they occupied when named more than two thousand years ago. Since sidereal hour angle is measured from the vernal equinox, and declination from the celestial equator, the coordinates of celestial bodies

> would be changing even if the bodies themselves were stationary. This westward motion of the equinoxes along the ecliptic is called precession of the equinoxes. The total amount, called general precession, is about 50.27 seconds per year (in 1975). It may be considered divided into two components: precession in right ascension (about 46.10 seconds per year) measured along the celestial equator, and precession in declination (about 20.04" per year) measured perpendicular to the celestial equator. The annual change in the coordinates of any given star, due to precession alone, depends upon its position; sunlight in summer and winter. Compare the surface covered by the same amount of sunlight on the two dates. Due to precession of the equinoxes, the celestial poles are slowly describing circles in the sky. The north celestial pole is moving closer to Polaris, which it will pass at a distance of approximately twenty-eight minutes about the year 2102. Following this, the polar distance will increase, and eventually other stars, in their turn, will become the Pole Star. The precession of the earth's axis is the result of gravitational forces exerted principally by the sun and moon on the earth's equatorial bulge. The spinning earth responds to these forces in the manner of a gyroscope. Regression of the nodes introduces certain irregularities known as nutation in the precessional motion.

The point to all this is to understand that mariners throughout history who navigated the globe using the stars recognized that celestial bodies are predictably moving targets. Adjustments needed to be made to calculations and tables to account for these movements. That is just on an earthly scale. Think about what that means to an extraterrestrial navigator traveling through the universe. Keep this in mind as we move forward.

Now consider the pivot point of the Eta Island line: Eta Island in Bermuda. Why would angels, ETs, or anybody put Eta Bermuda into play?

Angel Communication Code will not delve into the vast, dark hall of unsolved mysteries related to the Bermuda Triangle. We will look at the island from a purely geological perspective.

In May 2019, *National Geographic* published an article by Robin George Andrews titled "The Volcano That Built Bermuda Is Unlike Any Other on Earth." Without getting into a lengthy dissection of that article, I will present the opening introduction as published:

"No two volcanoes are the same, but they all form in the same handful of ways. All, it seems, except for the ancient volcano forming the foundations of the island of Bermuda. After examining rocks from deep under the island, scientists discovered that this quiet volcano formed in a way that is, so far, completely unique. The work, reported this week in the journal *Nature*, not only solves a long-standing mystery about this beautiful isle in the Atlantic, it also describes a whole new way to make a volcano."

For the sake of argument, we will accept the findings published by this reputable scientific journal. At a minimum, we can say that there is some level of uniqueness in the formation of the Bermuda Islands versus other volcanic islands. There may be a purely geological reason why Eta Island Bermuda is in play as a piece of the puzzle.

Then there is the infamous Bermuda Triangle as delineated by humans. The earliest suggestion of conspicuous and unusual disappearances in the Bermuda area appeared in a September 17, 1950, article published in the *Miami Herald* (Associated Press) by Edward Van Winkle Jones. Mysterious events in this region of the Atlantic Ocean are documented back as far as the days of Christopher Columbus in 1492.

Let us try to put this piece of the puzzle together. Recall that equilateral triangles were previously cited as one of the basic clues put before us that are leading us to a map for ET communication. The Bermuda Triangle is an equilateral triangle. If we put it on the (Google Earth) map along with the Eta Island lines, it looks like this:

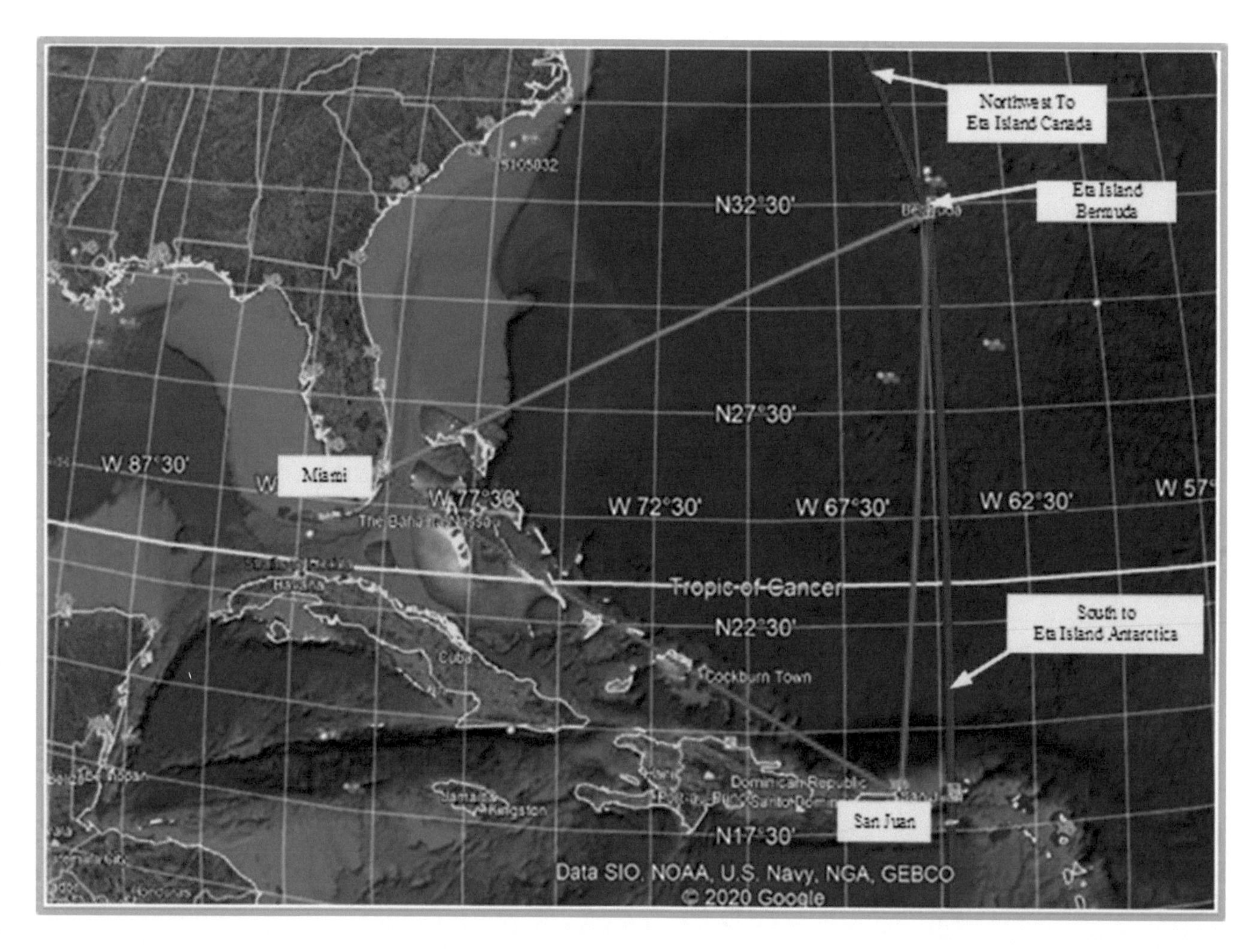

We can see the common point of the Bermuda triangle and the Eta line at Eta Island Bermuda. We can also see that the southern point of the triangle, San Juan, is only a couple of degrees off the Eta Island longitude line headed south toward Eta Island Antarctica. We also understand that a couple of degrees on earth are not much, and it is consistent with the margin of error discussed previously in relation to the location of the Arctic and Antarctic Circles. Consider that our delineation of the Bermuda Triangle may be off by a couple of degrees. Recall that the latitude of the northern tip of the Bermuda Triangle is also the northernmost latitude where Eta Carina Nebula can be seen in the Northern Hemisphere, as previously discussed.

The process we are following uses the number 3 as a driver, the first three prime numbers as the boundaries, and binary code as the language, and it has taken us this far. Now we are linking the elements of basic geometry to geography. We know that the earth is a dynamic environment because of plate tectonics and that the locations of geological features on earth are not permanently fixed forever. There has to be a reasonable margin of error allowed when undertaking any analysis of the location of things on earth over a period of thousands of years. What is in one spot today was not in that exact same spot forty million years ago. Precise locations of geological markers have changed over time, but they are close enough for practical use within the time window about which we are talking. The farther back in time we go with respect to the location of geological features on earth, the farther away those locations will be from today's locations.

With that in mind, it is reasonable for us to expect that ETs developing a map for us to find would have set it up such that the eastern vertices of their triangle would have fallen directly on top of the Eta line between Eta Island Bermuda and Eta Island Antarctica. After all, we arbitrarily decided the boundaries of the Bermuda Triangle based on actual events in a general region in our time in history and nothing else. It is not an exact science. We also know that things on earth move over time and that reasonable locational adjustments are required.

To put any adjustments to any location into a quantifiable perspective on an earthly scale, consider that the earth's circumference is approximately 24,901 miles. One degree of latitude is approximately 69 miles, or approximately 0.28 percent of the earth's circumference, which equals less than one-half of 1 percent. On an earthly scale, a few degrees of adjustment to the location of significant points translate into only a small percentage of the whole.

If we grab the Bermuda Triangle and hold the Eta Island Bermuda point, rotating it to make the eastern leg of the triangle align with the Eta line between Eta Bermuda and Eta Antarctica, then we find it looks something like this:

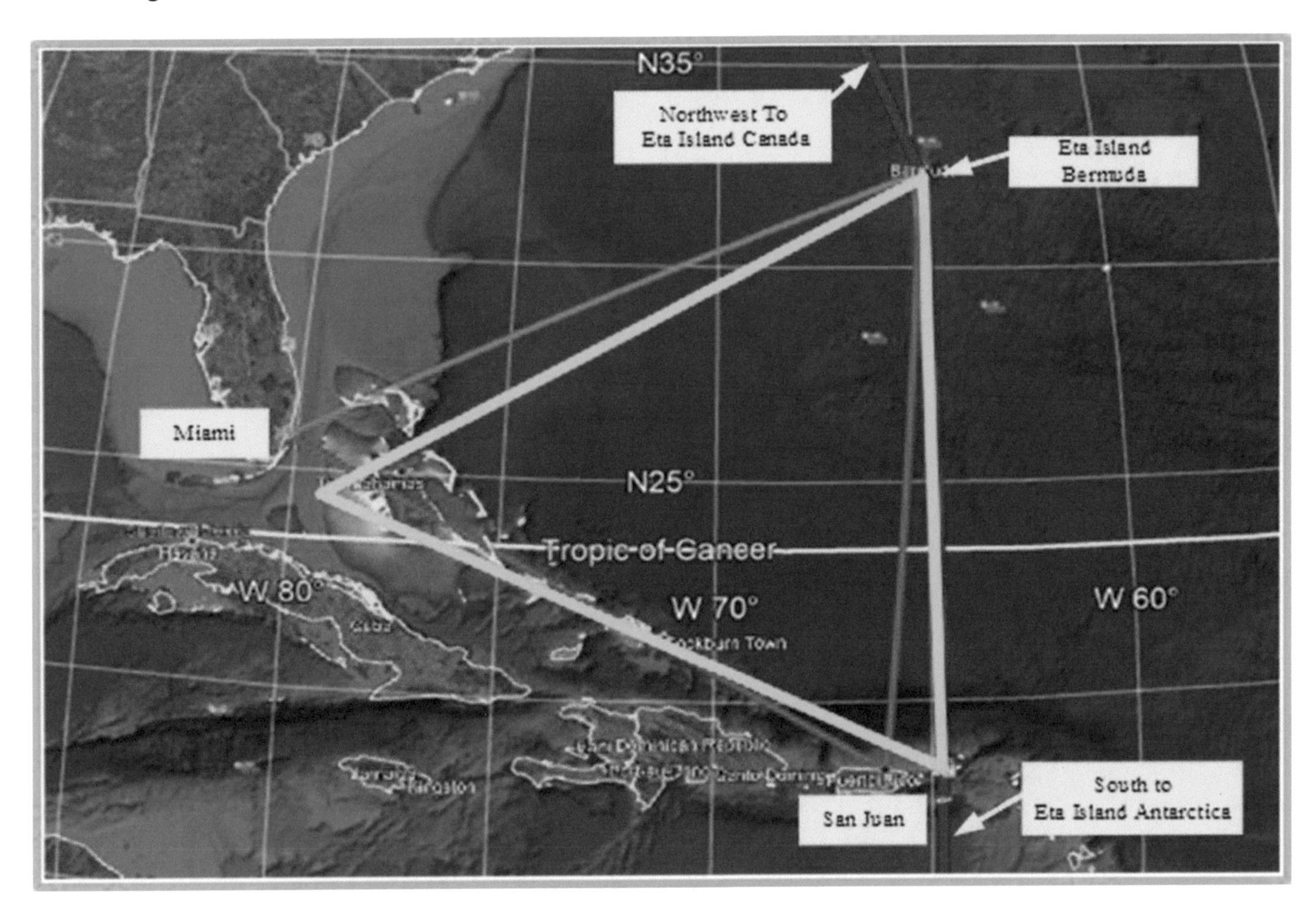

If we close the Eta Island lines and make another triangle by drawing the line that connects the Canadian Eta Island at the Arctic Circle and the Antarctic Eta Island at the Antarctic Circle, then it looks like this from outer space:

The closer you get, the more the adjustments of alignment show.

Up close, the rotated Bermuda Triangle and its proximity to the western leg of the Eta Triangle looks something like this:

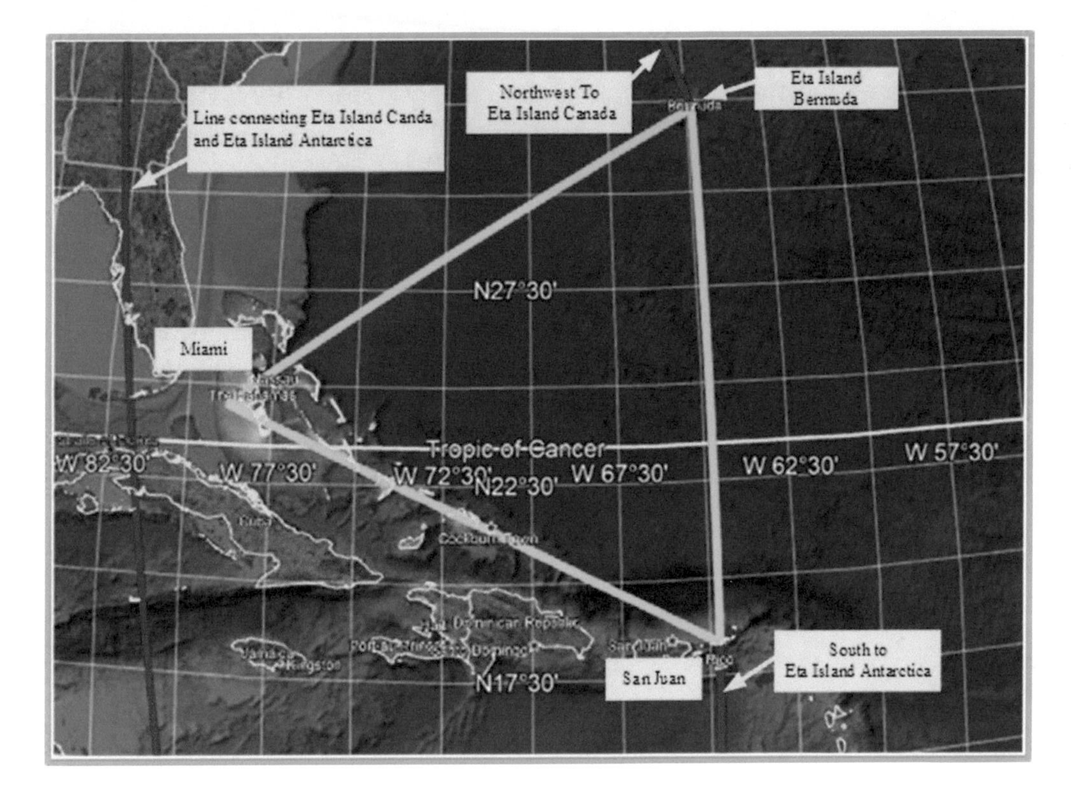

It would make some sense to posit that if ETs were developing some sort of map for us to find using control points to get a fix on position, they would have made the western point of the Eta Triangle lock down on the Eta Island triangle closing line that we show as passing near that point of the triangle. If we extrapolate the rotated triangle to achieve that closure and put all three versions of the triangle together, it looks like this:

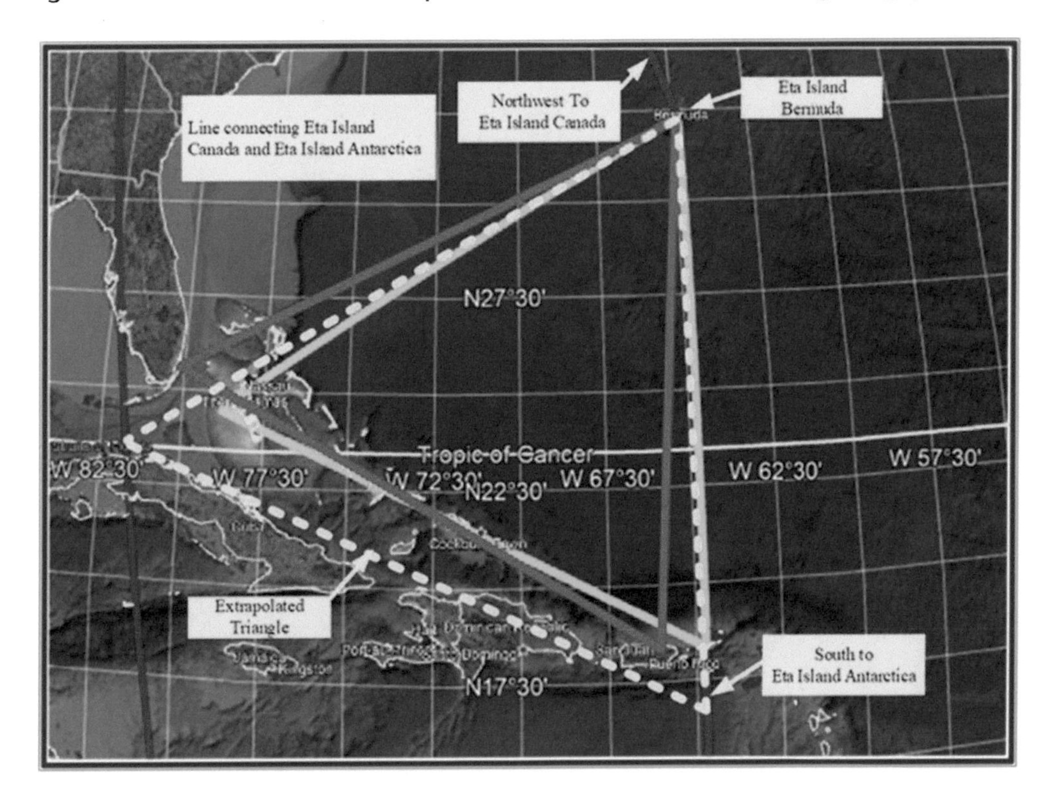

There is not much difference between the three interpretations of the triangle with respect to size and orientation on earth. Now the Tropic of Cancer latitude, the western point of the "extrapolated" Bermuda Triangle, and the western leg of the Eta Triangle all converge on a single point. The Tropic of Cancer is the northernmost point on earth where the sun can be observed directly overhead. During the summer solstice, the sun is directly overhead the Tropic of Cancer, and during the winter solstice, it is directly overhead the Tropic of Capricorn. The sun is directly overhead at high noon on the equator twice per year, at the two equinoxes. Spring (or vernal) equinox is usually March 20, and fall (or autumnal) equinox is usually September 22. Except at the equator, the equinoxes are the only dates with equal periods of daylight and darkness. At the equator, all days of the year have the same number of hours of light and dark.

Between the two tropic zones, which include the equator, the sun is directly overhead twice per year. Outside the tropic zones, whether to the south or the north, the sun is never directly overhead.[11]

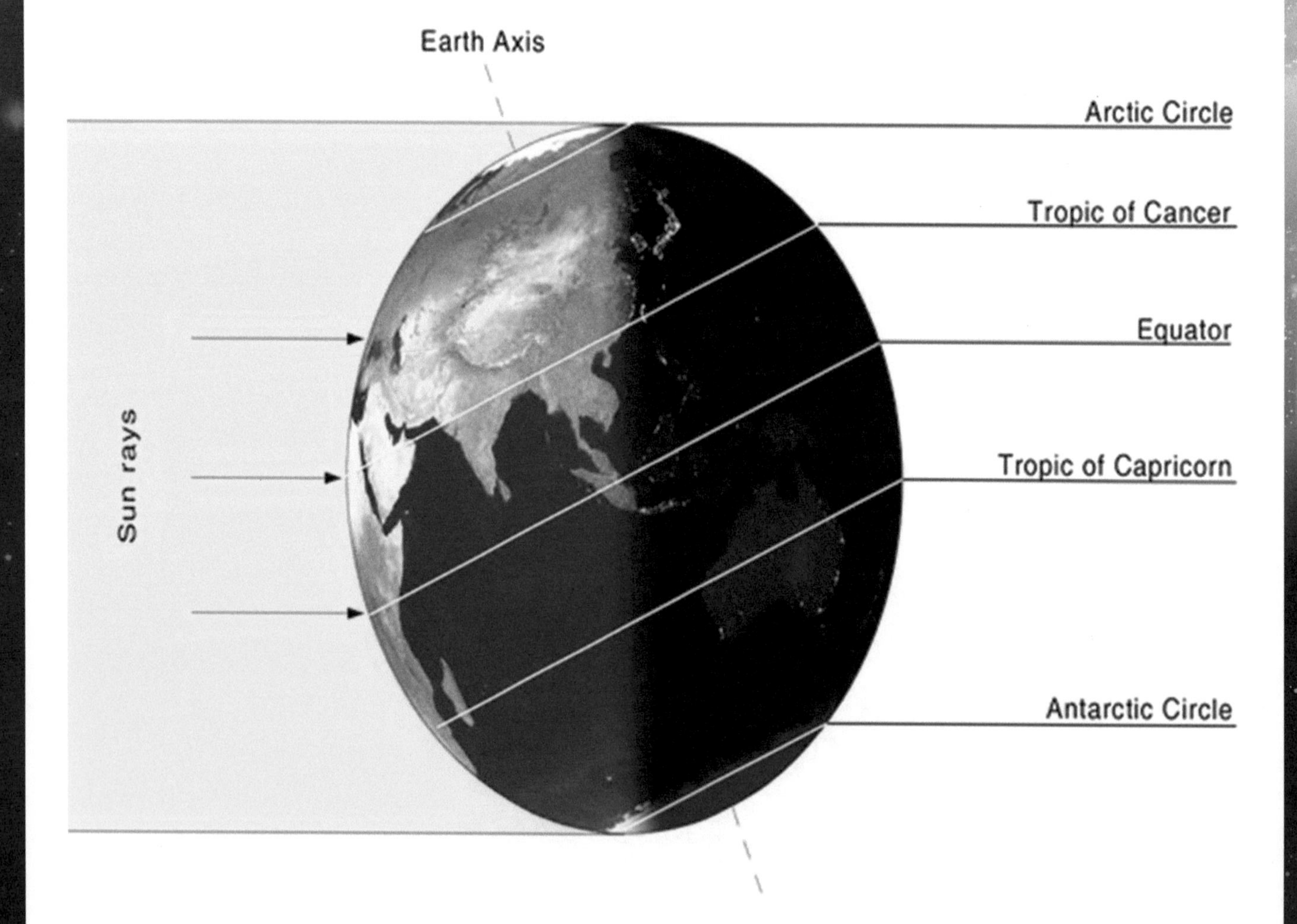

[11] "What Is a Solstice?," NOAA, https://scijinks.gov/solstice/, accessed November 24, 2023.

Winter solstice

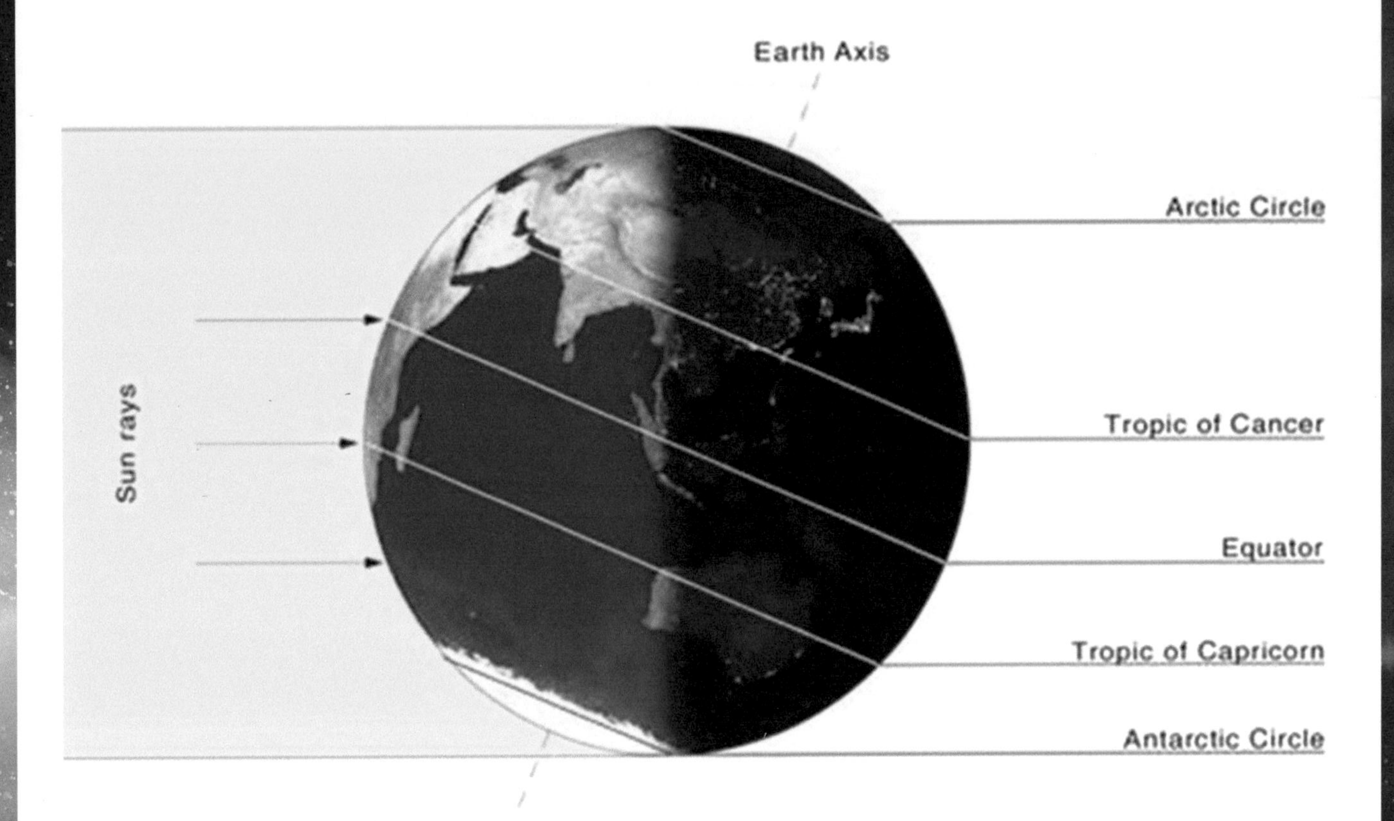

This means that within a reasonable margin of error of less than 3 percent in any direction, the Eta Triangle is locked down by three points on these three parallel earth circles:

1. the Arctic Circle—Eta Island Canada,
2. the Antarctic Circle—Eta Island Antarctica, and
3. the Tropic of Cancer—the western point of the Eta Triangle, which also falls on the line connecting Eta Canada and Eta Antarctica.

Much has been written about the Bermuda Triangle and all the unexplained phenomena that have happened within its boundaries. We are not trying to reopen that discussion here, although it seems that there may be a link to this place on earth (and those phenomena) and the results of our analysis so far. What is significant is that humans defined the boundaries of the Bermuda Triangle semiarbitrarily based on the frequency and nature of the many unexplained phenomena that have occurred within it. We, however, were brought to this location without considering any of those phenomena. We were brought to this location by the clues yielded from the binary model that created the trail for us to find and follow. The same result was obtained via two independent processes carried out by unconnected methods. Recall that we also had this sort of independent verification when we examined international Morse code earlier. The same result was obtained via two different methods carried out by independent parties. That is fairly strong evidence that there is merit to our methodology, and it has to mean something. At the very least, it lends more credibility than has ever been offered in the past to some of the celestial- and ET-based explanations of the mysteries of the Bermuda Triangle and gives the skeptics something much greater to consider.

As interesting as all this may be, our findings so far are nothing more than indicators that we are on the right trail to some bigger answers. We need to stay focused and on course to the end.

The three Eta Islands and the three Earth-specific latitude circles have essentially defined our triangle of interest as fixed on three earth circles. Now, let that triangle define a circle of interest to narrow our area of focus. For our purposes, we will continue to use the rotated and extrapolated triangle. It will become obvious that because of scale, whichever triangle is used does not make a significant impact on the eventual findings. One of the unique properties of the equilateral triangle is that its centroid and the center of the inscribed circle that it can hold are the same point. An inscribed circle within the triangle looks something like this:

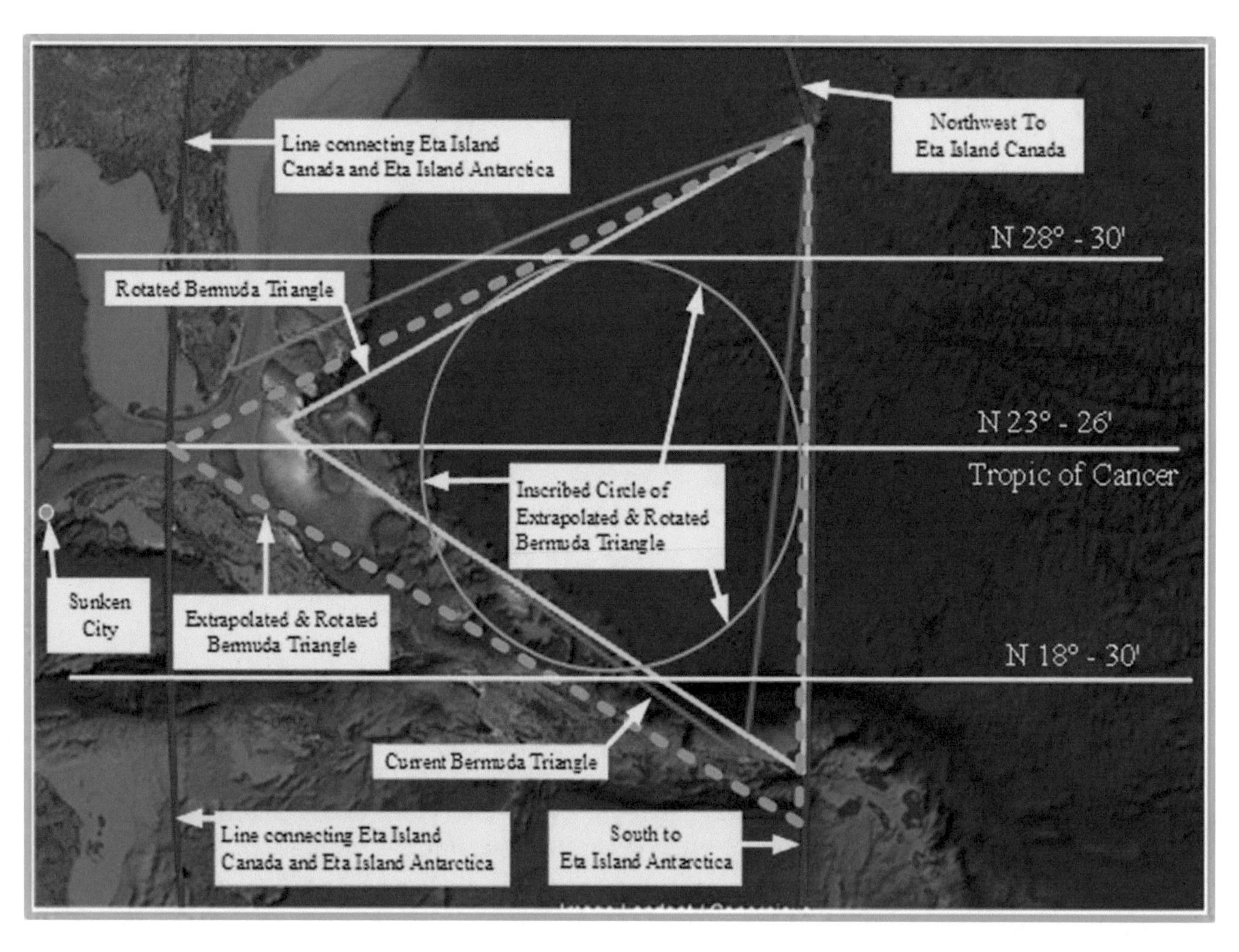

If we simplify all these lines to focus on the extrapolated triangle and the inscribed circle, it looks like this:

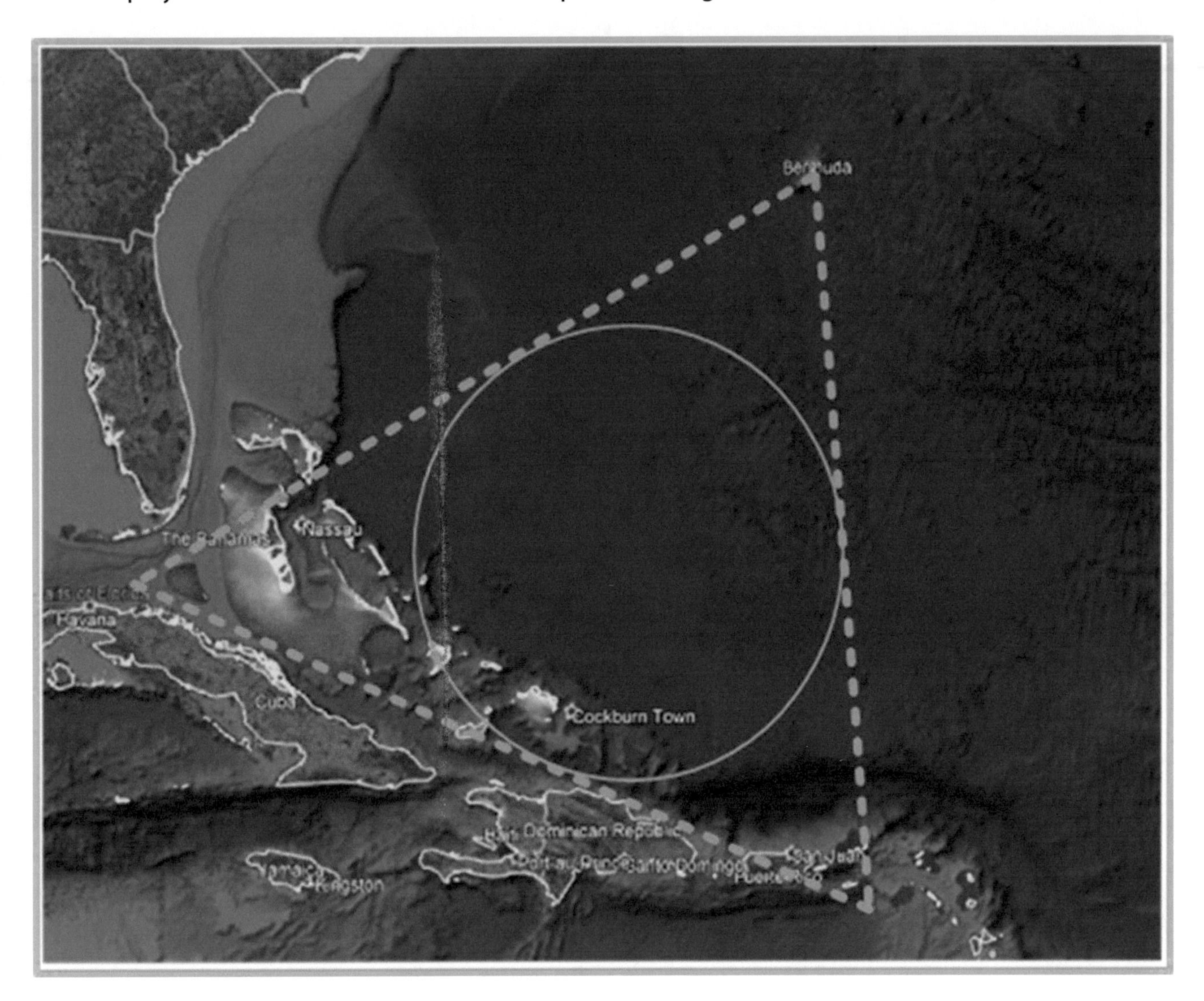

We are now brought back to the circle and triangle relationships previously discussed; however, this time it comes to us by way of actual places on earth and actual celestial bodies. We now have three triangles to consider, and that is the two Star of David triangles pointing north and south and this Eta Triangle pointing directly west in the direction in which celestial bodies fall below the horizon. If we add a duplicate triangle with a common core pointing directly east, where celestial bodies rise into view, then it brings the first sciences of astrology and astronomy back into play. It would build up like this:

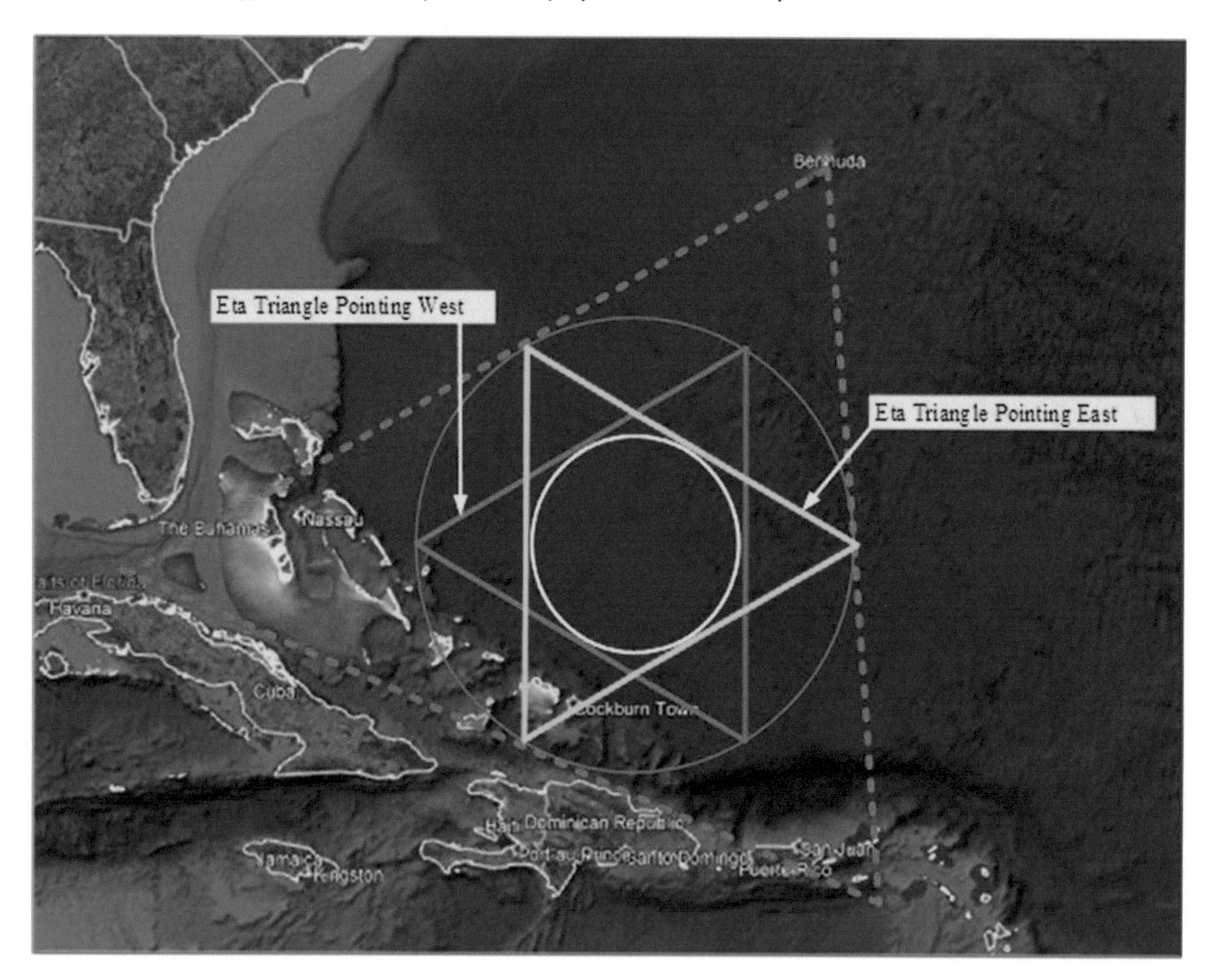

Now apply this to our zodiac calendar and it looks like this:

Amazingly, its uppermost points are directly at the transition point between the Era of Pieces and the Era of Aquarius, which is where we are at right now in time and in space. Recall our previous discussion of what the Age of Aquarius is all about. The Age of Aquarius is predicted to occur approximately 138 years from today. Aquarius is the sign of water. This Eta Circle clue is over water, with no landmass within it. The most essential ingredient of life as we know it is water. Aquarius is said to be the age when we will have our own renaissance, one that shifts toward humanism and science. Water indicates the nature of the coming Age of Aquarius, which highlights the qualities of genius, science, knowledge, humanitarianism, unity, and infinite possibilities. The triangles together are the Star of David rotated ninety degrees.

If we put the two sets of triangles together on the zodiac wheel, we see that it all comes together perfectly and looks like this:

There are twelve triangle vertices in this diagram. The number 12 is of enormous significance in the Bible. There are triangle vertices pointing to every era transition.

> The city had a massive high wall, with twelve gates. Twelve angels were at the gates; the names of the twelve tribes of Israel's sons were inscribed on the gates. There were three gates on the east, three gates on the north, three gates on the south, and three gates on the west. The city wall had twelve foundations, and the twelve names of the Lamb's twelve apostles were on the foundations. (Revelation 21:12)

The number 12 appears 187 times in the Bible, and 22 times just in the book Revelation, according to BibleStudy.org. Considered a perfect number, 12 symbolizes God's authority and also indicates as a perfect governmental foundation.

The number 12 also depicts completeness or the nation of Israel as a whole. For example, Jacob (Israel) had twelve sons, each of whom represented a tribe of the nation. This is important because if the New Jerusalem in fact will be populated by the whole people of God under the old and new covenants, then we have in this heavenly city the "true Israel" about which the New Testament authors write.

In addition:

- In Leviticus 24, God specifies that twelve unleavened cakes of bread be placed every week in the temple.
- In the Gospels, Jesus chooses twelve men as apostles. These men are to carry His authority after His ascension, to bear witness of His earthly ministry and finished work of redemption, and to take the good news to the world.
- In Revelation 7, we see 144,000 (12 × 12 × 1,000) sealed servants of God, and in Revelation 14 there are 144,000 people on Mount Zion with the Lamb.

The extensive use of the number 12 in Revelation 21 seems to illustrate that New Jerusalem represents "true Israel," or the redeemed people of God, throughout human history.[12]

There are seven archangels in the ancient history of the Judeo-Christian Bible; however, there are twelve archangels in the Jewish Kabbalah, whose names are Metatron; Raziel; Cassiel; Zadkiel; Camael; Michael; Uriel or Haniel; Raphael or Jophiel; Gabriel; and Sandalphon. According to the traditional Kabbalistic understanding, Kabbalah dates from Eden and its origins are traced back to the first man in Jewish cosmology, Adam. It is said that God revealed divine secrets to Adam such as the ten emanations of creation, the Godhead, the true nature of Adam and Eve, the Garden of Eden, and the tree of life.

The actual origins of Kabbalah are obscure, resulting from the fact that the practice was, for a long time, shrouded in secrecy amid closed circles, which restricted its study to only certain individuals, such as married men older than forty. These restrictions were imposed to preserve the tradition's secrets, which were considered too powerful, dangerous, and overwhelming to be handled lightly. Apocalyptic literature belonging to the pre-Christian centuries contained elements that carried over to later Kabbalah.

Early elements of Jewish mysticism can be found in the nonbiblical texts of the Dead Sea Scrolls, such as the Song of the Sabbath Sacrifice. The Bible provides plenty of material for Kabbalistic speculation, especially the story of Ezekiel and the chariot. The prophet Ezekiel's visions attracted much mystical speculation. In the book of Ezekiel, the prophet describes a somewhat mystical vision during which he sees very strange things such as wheels soaring through the sky or a valley of dry bones where the skeletons shake and rattle and suddenly reconstruct themselves into flesh and blood. Most importantly, the story of Ezekiel's encounter with God describes how the heavens open up and he sees four-faced figures emerge from a cloud of flashing fire: a man, a lion, an ox, and an eagle. Beneath their cloven feet, Ezekiel sees four wheels that move in conjunction with the figures, and he realizes the spirit of the four beings resides in the wheel.

---

[12] "The Number 12—Revelation 21:12–14," Once Delivered, January 31, 2017.

Finally, above the four figures, Ezekiel sees God sitting on a chariot or throne of blue lapis. The Lord gives Ezekiel His prophecies of doom and salvation for the Jewish people.

The unique nature of the book of Ezekiel caught the attention of the Kabbalists; no other prophets had written of their meeting with God in such mystical and vivid detail. Kabbalists believed that Ezekiel was recounting the realms that one passed through before hearing the voice of God. They reasoned Ezekiel knew that the age of prophecy was coming to an end and thus recorded his experiences so that future generations could continue on the same spiritual path.

The book of Ezekiel inspired much discussion on the mysteries of the heavens as the mystics pondered how they could progress on Ezekiel's path and achieve knowledge of God and the divine world. By studying the steps that Ezekiel described, the mystics believed they too could achieve divine prophecy and that anyone with the skills to reach God could find Him anywhere.

This was the era of early Jewish mysticism, which began sometime around the first century BCE. Other biblical sources of Kabbalah include Jacob's vision of the ladder to heaven and Moses's experience with the burning bush and his encounters with God on Mount Sinai.[13]

Kabbalists believe that God moves in mysterious ways. They also believe that true knowledge and understanding of that inner, mysterious process is obtainable and that, through that knowledge, the greatest intimacy with God can be attained.[14] One of many interpretations of the Kabbalah tree of life is shown below (a different sort of circle and triangle structure):

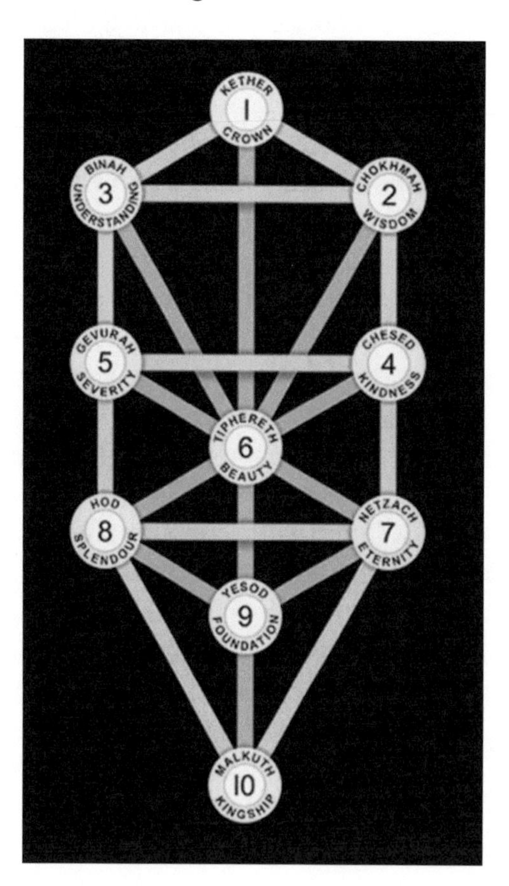

13 *New World Encyclopedia*, s.v. "Kabbalah."

14 Geoffrey W. Dennis, "What Is Kabbalah?," Reforming Judaism.

This journey we are on in *Angel Communication Code* is largely guided by a binary block "vision" and its first and last output: the word *eta*.

Getting back to the circle and triangle diagrams we are developing here in *Angel Communication Code*, let's see what it looks like when we put them all together in the same structure.

On the drawing of Metatron's Cube, it looks like this:

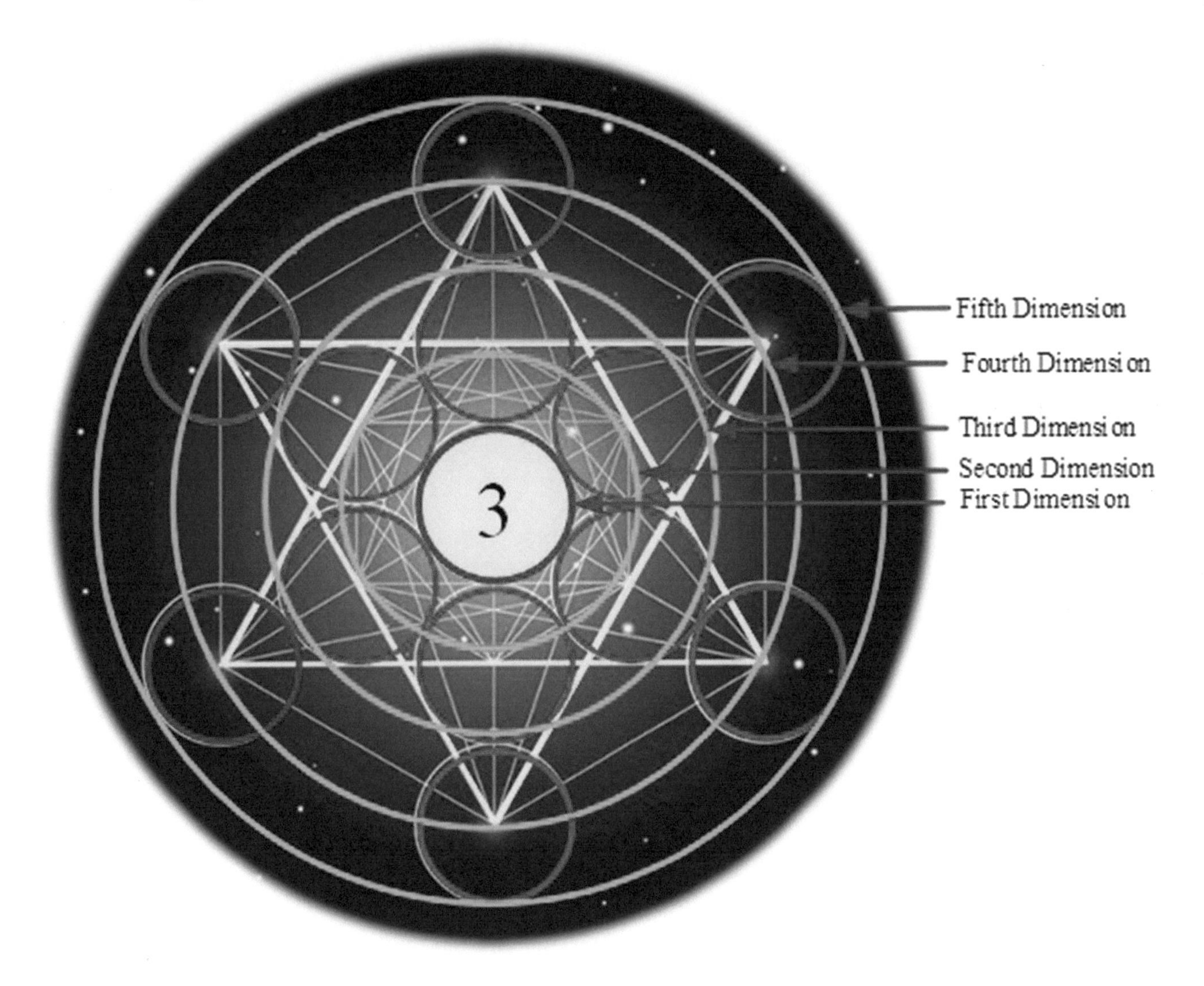

Add the two Eta Triangles and the Vitruvian Man and it looks like this:

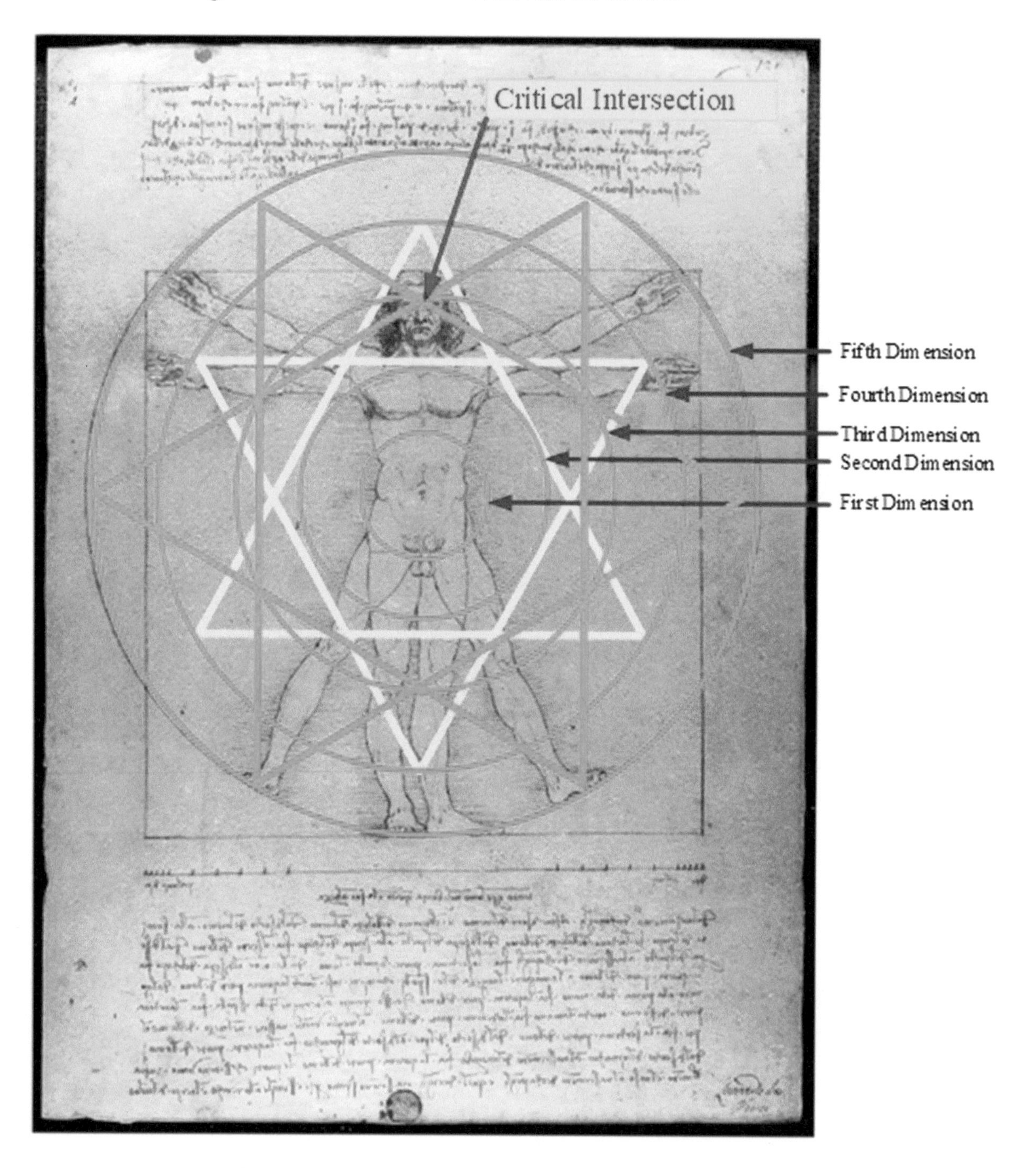

A critical intersection emerges at the location of the pineal gland in the human brain. This is where the limit of the third dimension of consciousness, the Eta Triangle pointing west, and the astrological triangle pointing east (where the constellations rise) all meet. That intersection is bound by the smaller upper triangle of the Star of David, pointing up toward the heavens. The diagram represents the universe in the human form, biblical/angelic form, and astrological form, in the first five dimensions of consciousness, in the human vibration frequency, and in the form of Archangel Metatron's Cube. That is a lot of universal energy directed to the center of that core circle, and that universal energy is what we seek.

In addition to all that energy, consider that these waters are influenced by the Gulf Stream and the Sargasso Sea and all that physical and biological energy. It is shown below with the Eta Circle superimposed:

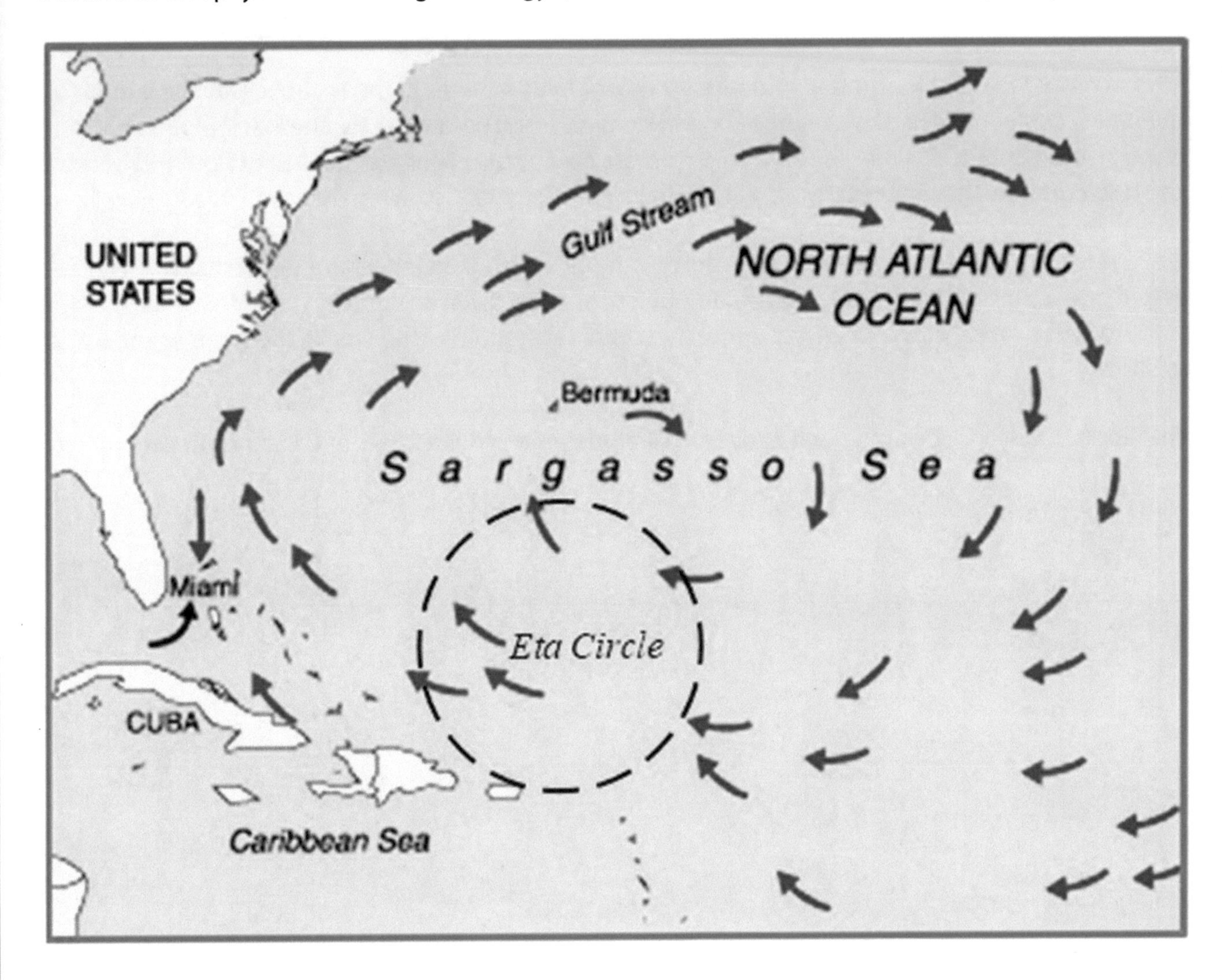

The Sargasso Sea is the only sea without a land boundary. An isolated area in the middle of the North Atlantic Ocean with our Eta Circle lying within it, the Sargasso is surrounded by the cold and turbulent waters of the North Atlantic, yet the sea itself is warm, clear blue, and calm. At the bottom of the sea is the Nares Abyssal Plain, which is about 3.6 miles deep. Various ocean currents swirl around the edges of the sea—on the western edge is the Gulf Stream; on the northern edge is the North Atlantic Current; on the eastern edge is the Canary Current; and on the southern edge is the North Atlantic Equatorial Current. These currents swirling around the perimeter of the sea create a calm spot in the middle of the ocean. The seaweed genus *Sargassum* is indigenous only to this sea. It grows thick and abundantly and can be seen floating on the surface of the water. For centuries, scientists thought the area was a vast desert with sparse nutrients, but today they have discovered the opposite. The currents that swirl around the sea feed into it, creating and sustaining many varieties of plant and animal life. The thick seaweed provides food and shelter for many sea creatures. We also now know that eddies and strong underwater currents that rise up carry nutrients from the bottom of the ocean to the top and give life to the sea. The Sargasso Sea is now compared to the rich teeming abundance of life of the rainforests.[15]

---

15 Sharleigh, "The Sargasso Sea—a Sea of Mystery," *10-4 Magazine*, Words of Wisdom, May 1, 2012.

An abyssal plain is an underwater plain deep on the ocean floor, usually found at depths between 9,800 and 19,700 feet. An abyssal plain is the result of the spreading of the seafloor (plate tectonics) and the melting of the lower oceanic crust.

Abyssal plains were not recognized as distinct physiographic features of the seafloor until the late 1940s, and until recently, none had been studied on a systematic basis. Because of darkness and a water pressure that can reach about 750 times the atmospheric pressure, abyssal plains are not well explored.[16] There is a lot of biological and geographical energy in and below this Sargasso Sea, which remains largely unexplored.

We seek a place from where to repeat back the message ETs and/or angels left for us to find, which guided us to this place on earth for that purpose. Our process has now given an address where that energy might come together on earth and when it could be most easily accessible, at the time of the summer and winter solstices.

That address is shown below, bounded by Earth latitudes and with its Tropic of Cancer centerline shown.

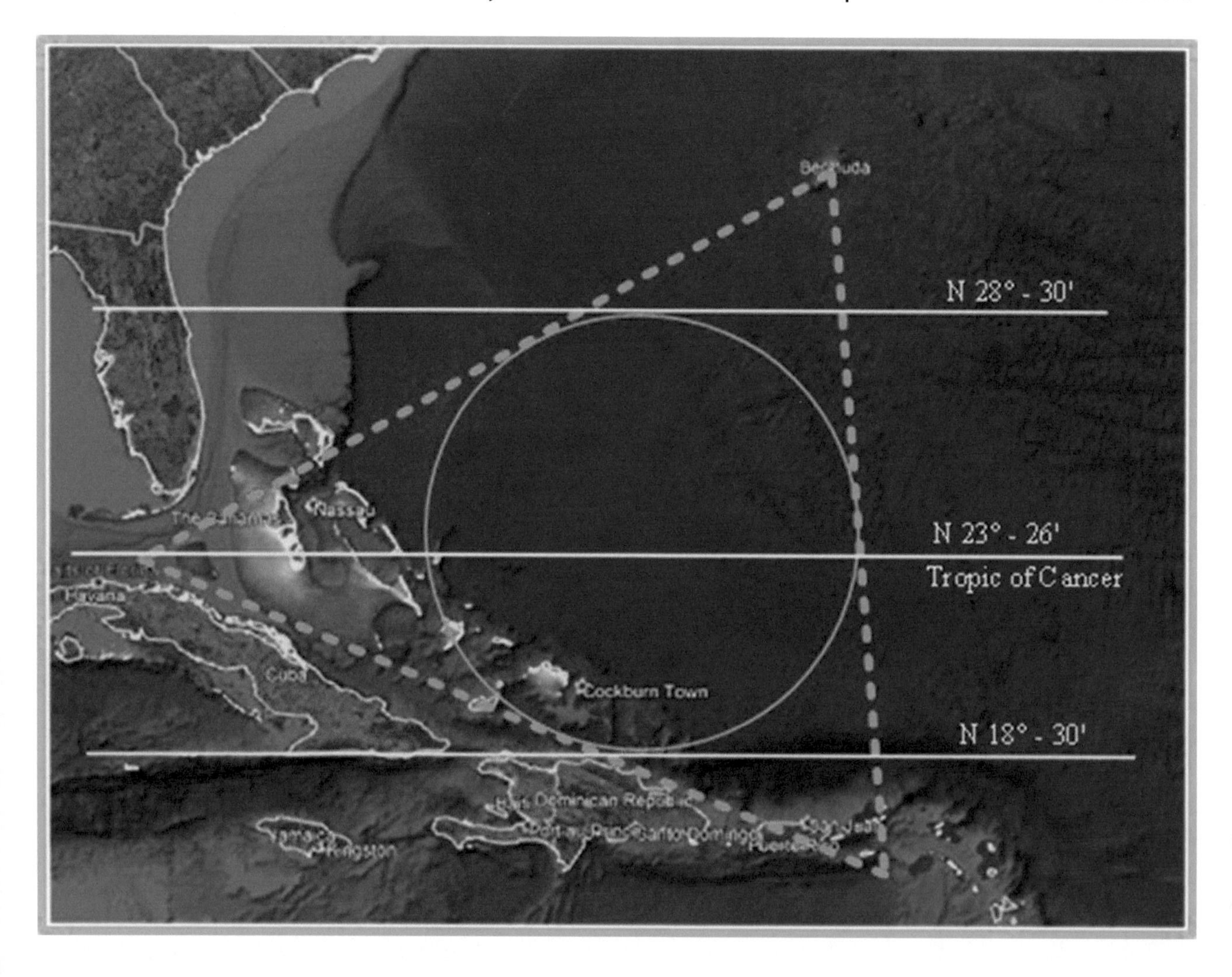

The distance between the upper latitude and the lower latitude is approximately 10 degrees or 690 miles. Note that the sum of our 2–3–5 binary code boundary condition is also (2 + 3 + 5) 10. This represents less than 3 percent of Earth's circumference, so it's a small target. Also, note that the Tropic of Cancer lies

16 Wikipedia, s.v. "Abyssal Plain."

approximately at latitude 23°26′ N. Twenty-six minutes is approximately thirty minutes (within a small percentage of error), or 0.5°. Therefore, the latitude of the Tropic of Cancer is at 23.5° N (2–3–5), which runs through the exact center of the Eta Circle. Recall that 23.5 spins per second is also the rotation of radio pulsar PSR J0250+5854. It is also the degree of tilt of Earth's axis.

Think about this for a minute.

a. 2–3–5 is a feature that describes the spin speed of the slowest known radio pulsar. The United States included pulsar information on our *Voyager* message in the 1970s.
b. 2–3–5 is a feature of the earth and its axial tilt angle.
c. 2–3–5 is the latitude of the Tropic of Cancer, which splits our Eta Circle right down the middle.
d. 2–3–5 adds up to 10, which is also the number of degrees of latitude bounding the Eta Circle.
e. 2–3–5 is the foundation of the binary block that leads us to these clues that describe communication, speed, angle, distance, and location.

This whole process started with a vision of the first three prime numbers, 2, 3, and 5. The circle and lines are not presented with pinpoint accuracy here, but they are presented with a very small and reasonable margin of error. We are slowly and methodically defining a fairly specific area of significance where we can continue with this process. Moving to a global, and then to a universal, scale will render any location-based discrepancies at this close-up scale irrelevant.

We have followed a long trail of clues to get to this Eta Circle of a specific size and in a particular location. We have arrived at the right location in our quest, but there is more work to do. We need to figure out how to use the information and the Eta Circle.

It seems that there are two ways to use this tool. The first is an "inside looking out" approach. Think of the Eta Circle as a spotlight or beacon pointing out into the universe with a fixed rotation. We can lock down its position on the earth right where we found it (on control points and lines) and let it rotate around with the earth on its 23.5° axis. It would be a pinpoint laser if that radial shaft of light (or radio signal) were projected out into the universe.

If we tracked that projection for one full revolution of the earth on its axis (one day), then the beam would delineate a specific area of space, and we could look at what the beam hit as it traveled around that circle. Perhaps it lands on specific targets of interest at particular times. It is reasonable, based on our discoveries, to suspect that the days we might want to focus on are as follows:

1. the vernal equinox in March
2. the summer solstice in June
3. the autumnal equinox in September
4. the winter solstice in December.

If we can somehow match this up with the locations in the universe where the view of Orion's belt from outer space is in some sort of alignment with our Eta Island line and determine when that alignment occurs, we will probably be getting very close to answering the questions of where ETs came from and where they go. Maybe this beacon shines on the location of the outer space end of the portal that connects to the Eta Circle–Earth end of the portal.

Our historic efforts to send out messages to the universe have focused on searching the sky randomly and trying to capture uniqueness in the twinkling of a light in the sky or in a new noise, then theorizing about what that twinkling or noise might mean. Of course, there is much more to it than that. The time has come to introduce a fresh approach to the search to get better results in terms of establishing ET communication. This fresh new approach should probably also consider the earth's precession.

Precession is the fluctuating axial tilt of the earth: the slow, but constant, gyration of the earth's axis. It takes approximately twenty-six thousand years to complete one full gyration. This earth dynamic was first discovered by Hipparchus in the second century BC. The simplest explanation is that the earth moves like a spinning top, so receiving and sending messages to and from ETs involves a moving target. A diagram of this dynamic is shown below:

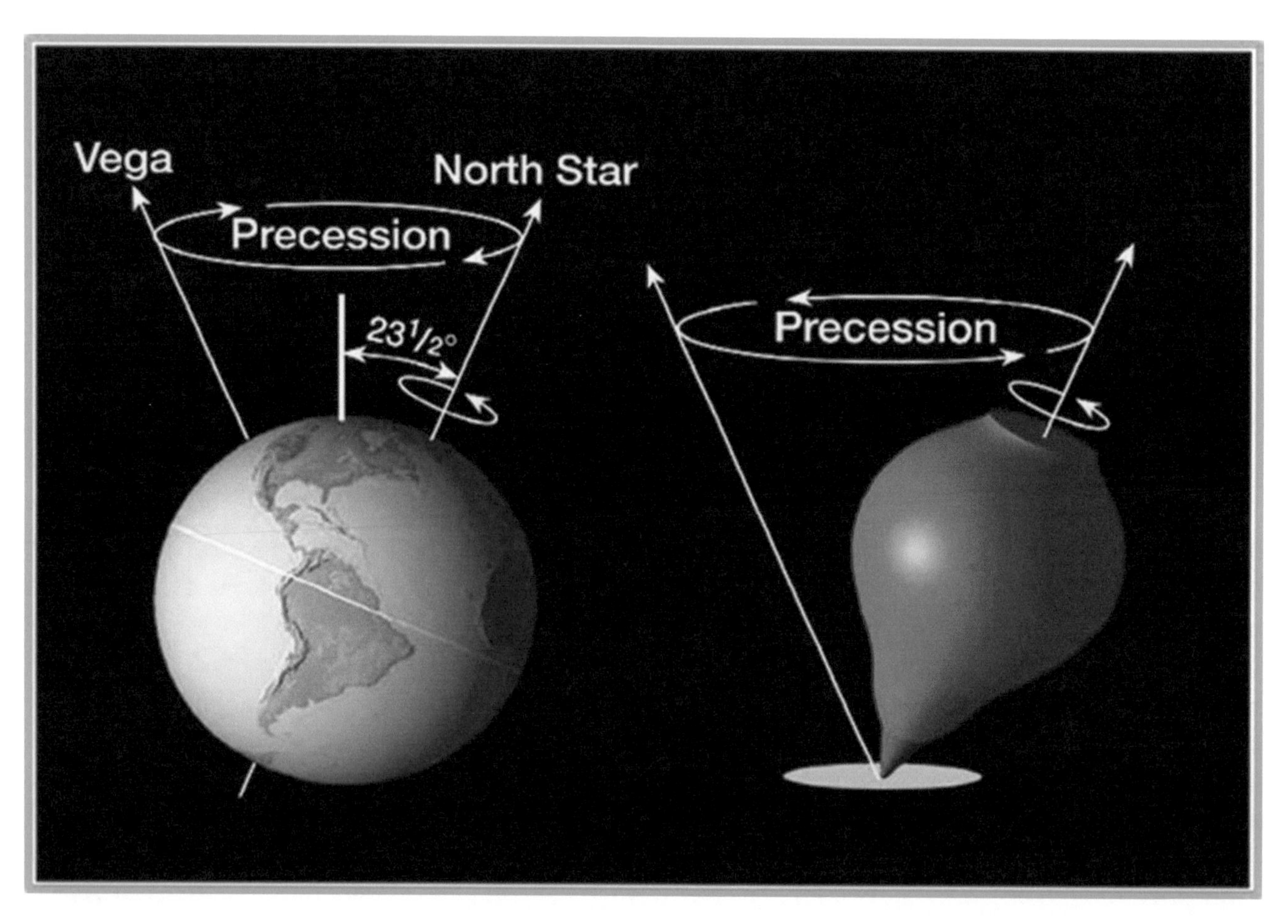

The possibilities certainly get one thinking differently. This dynamic may be involved in the identification of the direction in which to look out into the universe and narrow the search. It may help identify the portal that ETs use, which we found using ET-provided clues. This is a very different approach to a problem that has several moving parts.

The second approach might be to use the Eta Circle tool in an "outside looking in" approach. What happens if we lock down the Eta Circle within the Eta Triangle and let Earth rotate *under* it versus rotating *with* it? If we did that, it would track the shadow of the circle around the globe guided by the Eta Triangle's controlling latitudes of the Arctic and Antarctic Circles plus the Tropic of Cancer. We would see where on Earth this circle may have influenced humanity and perhaps where portals may historically have opened. We have to remember that ETs probably are (or were) not dialed into our man-made longitude and latitude lines. They probably used geological features and the sun as guides. Think of the Arctic Circle, the Antarctic Circle, and the Tropic of Cancer as tracks on which our Eta Triangle–inscribed circle could roll. The Eta Circle would traverse the globe, and we could see what it passed over. Putting the Eta Circle on parallel circle tracks would keep the Eta Circle locked between its upper- and lower-latitude lines. What goes on within these ten degrees of latitude around the globe?

The yellow band across the Earth map shown below represents the locations the Eta Circle would pass over as Earth rotated under it on its axis or the upper and lower latitudes that limit the circle. The numbered locations highlight numerous ancient structures of significance that lie within this band of latitude. The point is to demonstrate that the band goes right through the site occupied by the Mayan civilization and that of the Egyptian civilization. It also covers Mexico's Zone of Silence (item 1 on the map). These places were/are home to some of the most significant ancient civilizations and structures currently known to humanity, not excluding the Bermuda Triangle. The width of this band is a small percentage of Earth's latitudinal circumference. There are a lot of significant ancient sites in a relatively small area.

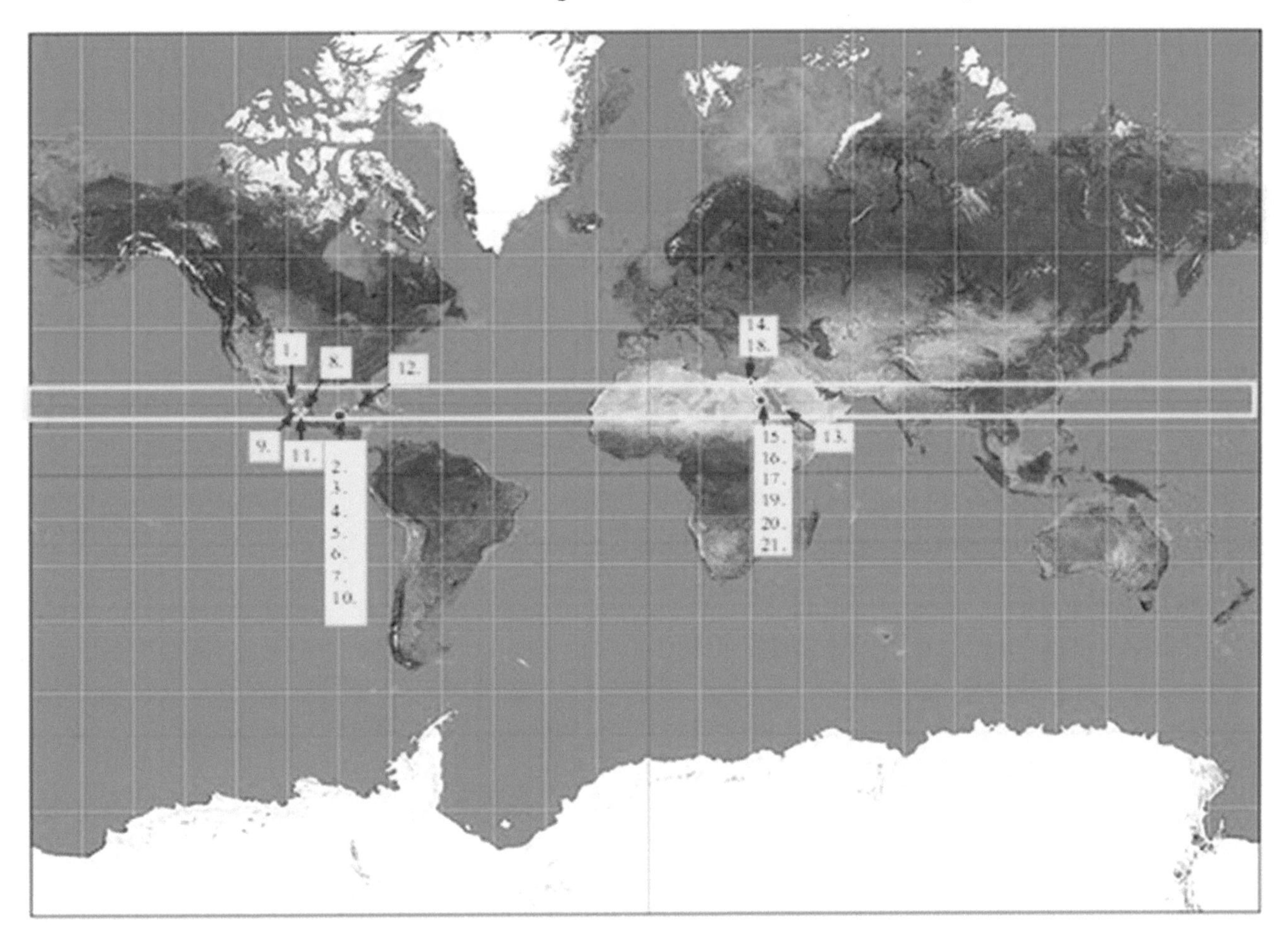

The numbered locations on the map correspond to the following ancient civilization sites:

1. Mexico's Zone of Silence
2. Chichén Itzá
3. Tulum
4. Ek Balam
5. Coba
6. Uxmal

7. El Rey
8. El Tajin
9. Tula
10. Labna
11. Templo Mayor
12. Cuba Sunken City
13. Mecca
14. Pyramids of Giza
15. Temple of Karnack
16. Abu Simbel Temple Complex
17. Valley of the Kings
18. Saqqara Pyramids
19. Luxor Temple
20. Temple of Hatshepsut at Deir el Bahari
21. Midinet Habu.

Several of these ancient locations have structures that align with either Orion's belt or solstice/equinox events. It has been speculated that several of these sites may be the location of portals leading to the far reaches of the universe or to other dimensions of consciousness.

What is important here is that we are now driven to this observation by way of constructing and then interpreting a block of binary code. We were led to clues that may have been provided by ETs and/or angels and to the first three prime numbers.

It appears so far that the word *eta* generated by our analysis is sending a message about location and communication: the location of important and relevant things on earth, and the location of potential search targets in the universe. It also suggests the location of the Earth end of a wormhole that we need to access in order to establish communication.

It seems very plausible now, based on some evidence, that the Eta Circle might hold the energy necessary to open a wormhole-like portal at specific times of the year when the alignment of Orion's belt is just right. We were guided to eta through a process. It was the first word and the last number revealed in that process, and clearly it has a lot of meaning related to the premise of *Angel Communication Code*.

CHAPTER

8

Message Implementation

It has been stated in *Angel Communication Code* that:

a. A message (at least one) was provided to humans by extraterrestrials and/or angels that will lead humans to a way to establish two-way ET communication.
b. The key to decoding the message was not provided in an obvious manner. It had to be figured out, requiring that humans demonstrate a sufficient level of advancement before ETs will openly communicate with them.
c. The key to the coded message was found in a dreamlike vision, like many other important scientific discoveries in human history.
d. In order to make the ET connection, we must find and respond to the message we were provided in a "repeat-back" approach. It is a very basic and natural process
e. Our historic efforts to send original messages out into the cosmos have yielded no results, most likely because—
    - The messages we composed are too complicated to decipher.
    - The messages were sent without proper consideration of content and without specific instructions as to when or where to send a return message.
    - Our approach to sending messages into the cosmos over the years did not follow the fundamental scientific method. We have essentially repeated the same process expecting a different result. No substantial adjustments were made to the original hypothesis or the original experiments. Again, we have not properly followed the basic scientific method.
f. The deciphering of the message leads us down a trail of numbers that are exposed and surface in different forums, yet they are conspicuously linked in many unmistakable ways.
g. Gathering up all the numeric and other information and putting it all together leads us to a specific location that may hold the energy needed to get our repeat-back response where it needs to go to be heard, if transmitted at the right celestial time or times of the year with respect to the position of the sun.

All this information so far leads us to the development of a hypothesis consistent with standard and proven scientific methods.

The only way anyone can ultimately prove or disprove the observations and concepts presented in *Angel Communication Code* is to design a good experiment, execute the experiment, and study the results. This is the basic premise of the scientific method. It makes sense to try given that what has been done historically regarding ET communication has not achieved the desired result of establishing communication with ETs.

Billions of dollars have been spent seeking the solution to the ET communication problem. SETI alone has an operating budget of approximately $2.6 million per year as published on their year 2020 tax form 990, which can be found at https://www.seti.org/about-us/financials. Then there are the billions of dollars spent by NASA and other global government and private entities. None of all this money has established two-way ET communication as far as the general public knows.

It has all gotten too complicated, yet the means and methods do not change, nor do the results. This is why we must go back to basics, begin relying on the fundamental scientific method, and stop wasting money on what we know has not worked thus far.

Seth Shostak, senior astronomer and leader of the Search for Extraterrestrial Intelligence Institute, or SETI, gave the talk "When Will We Find E.T. and What Happens if We Do?" at NASA's Langley Research Center on August 8, 2017. Part of that lecture was about the nature of a hypothesis:

> **Exploring a Hypothesis**
>
> Even so, the overall search effort is accelerating, he said. "The speed of the search is going up with time," Shostak said. Digital innovation will continue to produce increasingly powerful tools. "In almost every year, you can say that we're collecting more data this year than in all previous years put together," he said. "That's usually the case."
>
> The lack of any evidence to date has not slowed SETI's efforts. Shostak noted that the group's work is different from other scientific projects in which researchers make a hypothesis, test it, and—if the evidence fails to support the premise—move on to a new theory.
>
> "SETI is not like that," Shostak said. "The hypothesis is that there's somebody out there as clever as the average resident of Virginia Beach—but there's no way to disprove that. You can't prove they're not there."
>
> He compared the search for extraterrestrial beings to exploratory expeditions, like those that once set out in search of Antarctica.
>
> "It's like hypothesizing that, hey, there's a big continent at the bottom of the world," Shostak said. "You could argue about it forever … but in the end, you have to send some ships down there and find out. That's what SETI's all about."

Science is exponentially advancing in our modern age, and there is a much more concentrated effort being put into establishing extraterrestrial communication beyond the observance and recognition of UFOs. In *Angel Communication Code*, we are conceding the fact that ETs do exist, they have been coming and going for thousands of years, and they have had an influence on the development of the human animal for millennia. Thus far, we have looked at the evidence and identified many observations that warrant the revised hypothesis proposed in *Angel Communication Code*, just as Shostak recognizes in his lecture at NASA's Langley Research Center. His key statement as far as *Angel Communication Code* is concerned is this: "If the evidence fails to support the premise—move on to a new theory."

We know that the scientific community has not produced the desired results after decades of research, experimentation, and the expenditure of billions of dollars. That is astonishing, yet the ET search continues, although the means and methods of continued experiments are generally the same. The only thing that seems to change is the advancement of the tools available for application.

There are various definitions of the scientific method, but the fundamental process is as follows:

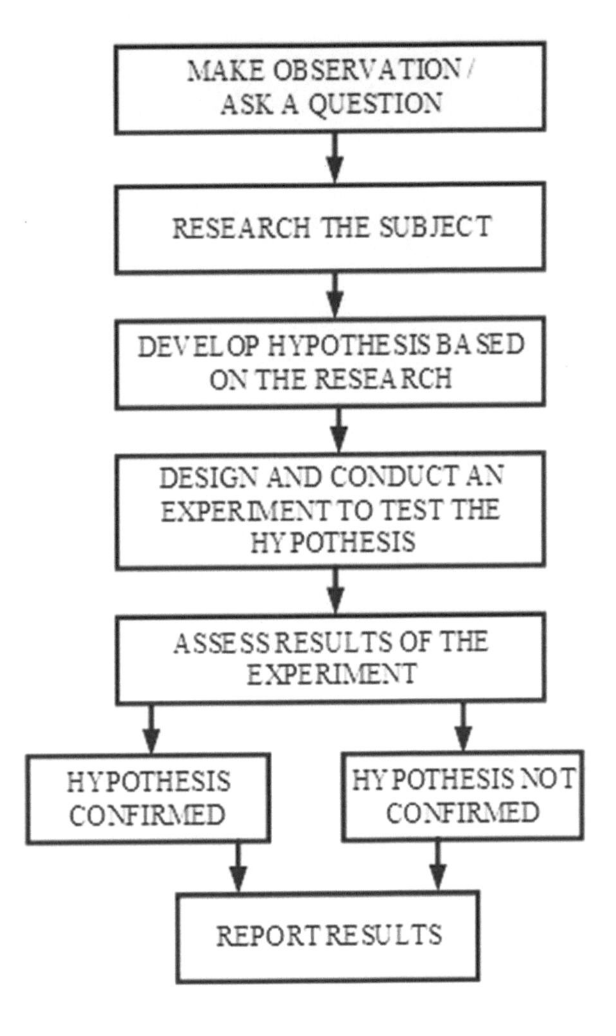

Angel Communication Code suggests that going back to basics is the right thing to do, as is so often the case in many situations. Let us start from the beginning and redevelop a hypothesis based on what we have learned.

1. Make an observation / ask a question.
   Some of the key observations made to date with regard to communication with ETs are:

   a. According to numerous pieces of evidence, including eyewitness accounts and video accounts, ETs do in fact exist.
   b. Documentation exists from important ancient texts through writings of modern times that ETs have contacted individuals and entire cultures on earth.
   c. Ancient texts have reported that ETs have either created or influenced those cultures.
   d. We have attempted to send coded messages into the universe but have received no documented response from ETs to those messages. It's been going on for decades.
   e. Modern science believes that ETs exist and has spent billions of dollars trying to establish contact—and continues to do so. The budgets exist to try a new approach.

f. Modern science believes that wormholes exist and that their existence can be proven based on the known laws of physics; however, a wormhole has not been found to date.
g. The observations described in *Angel Communication Code* are based on fact and lead us to a different approach to establishing ET contact, different from the unsuccessful approaches used in the past.

2. Ask questions about the observations (cited above) and gather information.
   Are the observations listed above accurate?

   a. Yes—there is too much consistent physical evidence and too many eyewitness accounts in the modern era, from all over the world, to conclude that all this ET information is false.
   b. Yes—there are too many accounts throughout history from all over the world to conclude that the written accounts are all false.
   c. Yes—there is reasonable evidence to suggest that ETs have influenced human culture.
   d. Yes—to the best of our knowledge, no contemporary ET communication has been received.
   e. Yes—the search for communication signals from intelligent life in the universe continues at great expense and with no results.
   f. Yes—modern astrophysicists agree that wormholes must exist, but nobody has found one yet.
   g. Yes—the suppositions made in *Angel Communication Code* are based on facts, and those facts lead us to the conclusion that an experiment is warranted based on a credible hypothesis.

3. Form a hypothesis.
   A hypothesis is an explanation of the observations and the formulation of predictions based on those observations.

   ET communication hypothesis: It is possible that our attempts to establish two-way communication with ETs have received no response because of the following reasons:

   a. We are sending the wrong messages in hopes of receiving a response. The messages are too complicated to decipher and are not consistent with three-way/repeat-back communication procedures.
   b. We are sending our messages in the wrong direction. We have not been able to target the wormhole necessary to close the time gap and get our messages to the area where we believe they might be received in a timely manner. We have essentially used a "message in a bottle" approach. We put our messages either in a ship (satellite) or on the waves (radio) and launched them with insufficient consideration of specific direction—just hoping an ET finds it.
   c. We are sending our messages at the wrong times. We need to send them when a portal is open. The wormhole/portal might be a predictable, but moving, target. It is possible that the opening of wormholes is influenced by equinox and solstice events here on earth.
   d. The information presented in *Angel Communication Code* is sufficient to warrant an experiment in that it:
      i. identifies the correct message format and provides a rationale,
      ii. identifies a location from where to send the message that has a higher probability of success, again with a rationale behind it, and
      iii. identifies a time to send the message that has a higher probability of success.

e. If the correct message is sent from the correct location and at the correct time, we can expect to receive a response, consistent with three-way communication procedures, in a timely manner.

4. Test the hypothesis with an experiment that can be reproduced.
   a. In the next section of *Angel Communication Code*, an experiment is proposed that can be reproduced.
   b. The experiment would be reasonably economical compared to the investments already made that have not produced the desired results.
   c. The experiment is designed in a way that has not yet been tested. We would not be repeating a process that so far has not produced the desired result.

5. Analyze the data and draw conclusions.
   a. Accept or reject the hypothesis based on the data received from the experiment.
   b. Modify the hypothesis and the experiment if necessary.

6. Reproduce/repeat the experiment with modifications as necessary.
   That comes after we test this new hypothesis.

Now let us revisit and expand the fundamental scientific method model based on what we have learned so far by experimentation and the results achieved to date regarding the establishment of two-way ET communication.

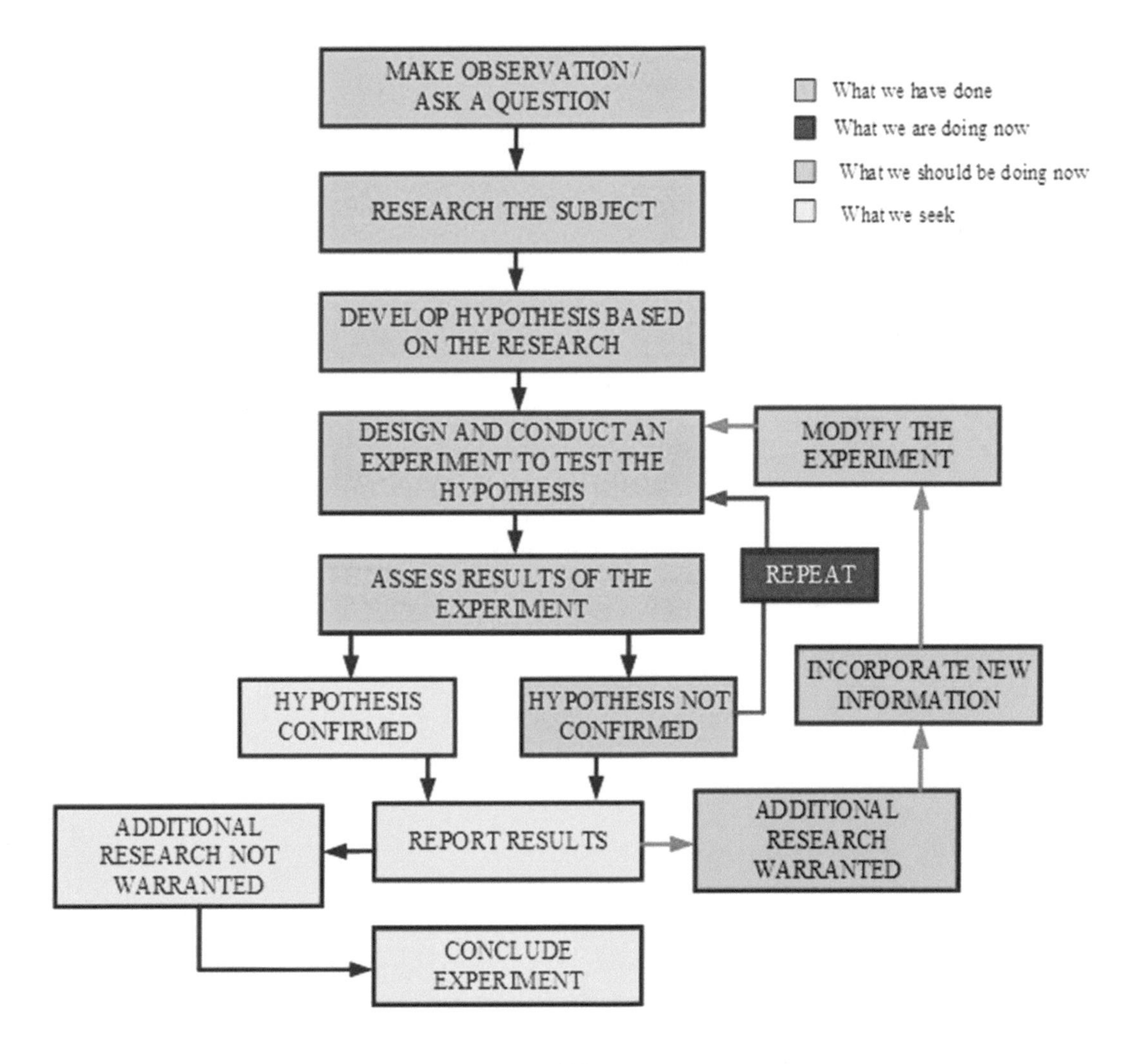

Consistent with the scientific method, we need to be assessing what we have done, which has not produced the expected result, and revising the experiment with new information and a new approach to the problem. The alternatives are either to abort our efforts to establish two-way communication with ETs or to keep repeating the same process hoping for a different result. It seems that the world is not satisfied with the results so far, nor is it willing to concede. The appropriate course of action now becomes obvious in this light: change the approach and continue the search. A different approach, based on unique new observations, is what we present in *Angel Communication Code*.

Continuation means:

a. acknowledging and accepting that what we have been doing to date has not produced the expected result and
b. recognizing that it is a distinct possibility, based on plausible evidence, that our approach needs to be a repeat-back of a message that was provided, versus trying to conjure up another original, complex message that we hope will be found, deciphered, and understood by our extraterrestrial brothers and sisters.

To accomplish this, we need to figure out what the ET message is and how exactly to repeat it back. In the book *Extraterrestrial Communication Code* referenced throughout *Angel Communication Code*, we developed a repeat-back message built around the prime numbers 2, 3, 5, and 7 and four words that fell out of the binary block that was derived from a vision/dream (Eta, Pas, Swy, and Zac). In *Angel Communication Code*, we will create a different repeat-back message that is even simpler and more focused on the word *eta*. *Eta* is a three-letter word that seems to offer the most meaning to what we are discovering in that:

1. Eta is the first word that dropped out of our code analysis, corresponding to the first prime number, 2.
2. Eta is the only prime number produced from our code analysis using only prime numbers 2, 3, and 5. Recall that *eta* is ancient Greek for the number 7 and also is the seventh letter of the ancient Greek alphabet. Seven is the fourth prime number.
3. The letters that make up the word *eta*, as noted previously, are the most frequently used letters in the English alphabet, as discovered by Samuel Morse in the development of his Morse code, which was the first electronic communication code ever invented.
4. The word *eta* describes many relevant things such as eta-Earth, eta function, and celestial bodies in the universe linked to the Albert Einstein equations used in the search for wormholes / black holes, time portals in the universe, and more. This word seems to be the beacon of light in our quest.
5. Eta is the name of three, and only three, islands on this earth, which, when examined together, are linked to the Bermuda Triangle and numerous ancient megalithic structures all over the world that all have ties to each other.
6. The Eta Circle derived from the Eta Triangle is unique in that it is not a physical structure chosen by beings but, rather, is a geographic location, one to which the analysis of our code led us directly and conspicuously. In addition, *Angel Communication Code* makes links to the Eta Circle, Da Vinci's Vitruvian Man, Archangel Metatron's Cube, ancient astrology and horoscopes, the human vibration frequency, dimensions of consciousness, and more. There is a lot going on in the word *eta*. There seems to be a lot of earthly and cosmic energy in this Eta Circle, which *Angel Communication Code* has exposed and linked together for the first time ever.

The message we propose therefore is a repeat-back of the word *eta* in a specific and recognizable pattern and from a specific location on Earth. That location is the Eta Circle within the Bermuda Triangle longitudes as shown in *Angel Communication Code*. That is the starting point.

Building the code with only the word *eta* starts with binary code. Note that the *E* and *T* are capital and the *a* is lowercase, with capital and lowercase letters having different binary values. This is the way it came out of the code analysis. Recall the following:

	Binary	**Morse code**
E =	01000101	•
T =	01110100	–
a =	01100001	• –

A repeat-back message designed around the word *eta* and the prime numbers in our code, 2, 3, 5, and also 7, is shown below. It uses both binary and Morse code.

1

Day Before Equinox / Solstice					
Pattern			Seconds	Binary	Morse
E		1	3	01000101	•
		2	7	Gap	
T	1	3	3	01110100	–
		4	7	Gap	
a		5	3	01100001	• –
			7	Gap	
E		1	3	01000101	•
		2	7	Gap	
T	2	3	3	01110100	–
		4	7	Gap	
a		5	3	01100001	• –
			7	Gap	
E		1	3	01000101	•
		2	7	Gap	
T	3	3	3	01110100	–
		4	7	Gap	
a		5	3	01100001	• –
			7	Gap	
Total= 90 Seconds					

2

Day of Equinox / Solstice					
Pattern			Seconds	Binary	Morse
E		1	3	01000101	•
		2	7	Gap	
T	1	3	3	01110100	–
		4	7	Gap	
a		5	3	01100001	• –
			7	Gap	
E		1	3	01000101	•
		2	7	Gap	
T	2	3	3	01110100	–
		4	7	Gap	
a		5	3	01100001	• –
			7	Gap	
E		1	3	01000101	•
		2	7	Gap	
T	3	3	3	01110100	–
		4	7	Gap	
a		5	3	01100001	• –
			7	Gap	
Total = 90 Seconds					

3

Day After Equinox / Solstice					
Pattern			Seconds	Binary	Morse
E		1	3	01000101	•
		2	7	Gap	
T	1	3	3	01110100	–
		4	7	Gap	
a		5	3	01100001	• –
			7	Gap	
E		1	3	01000101	•
		2	7	Gap	
T	2	3	3	01110100	–
		4	7	Gap	
a		5	3	01100001	• –
			7	Gap	
E		1	3	01000101	•
		2	7	Gap	
T	3	3	3	01110100	–
		4	7	Gap	
a		5	3	01100001	• –
			7	Gap	
Total = 90 Seconds					

There is a variety of approaches to the transmission of this information. In this version, each iteration has a transmission time of ninety seconds each for binary and Morse. There are numerous ways to approach this; we will leave it to the reader to explore other options. The presented approach would be to transmit E–T–a in binary, then follow it with the same in Morse. It is a simple and easily recognizable pattern.

Given that there are 864,000 seconds in a day, we could complete 9,600 cycles in a day. This would be 28,800 cycles of a very simple pattern over the three-day event. That would get some attention if received for sure, and it'd be far less complicated than what we have tried in the past, and what we continue to do with respect to the design of an appropriate message to launch into the universe. It seems more likely to succeed if the message is familiar and simple, versus trying to send some sort of code explaining complex concepts such as the structure of human DNA. We are not trying to educate the ETs, just trying to say hello. We must start the conversation by saying hello using familiar codes and patterns that have been provided and used throughout our history—a simple and recognizable code pattern delivered from the right location at the right time. We need to start by saying hello back to the ETs' original hello, and not by trying to provide them a PhD-level education placed on a golden record album that can only be played if they themselves are able to follow the instructions we provided for constructing a record player. That approach has proven to be unsuccessful. This new approach is worthy of an attempt, consistent with the scientific method, based on the discovery of new information found through research conducted using a different way of thinking about the problem.

CHAPTER

9

Proposed Experiment

The proposed means and methods for seeking and sending a message to, and receiving messages from, ETs are to put the equipment on an oceangoing vessel specifically outfitted for this purpose, then stationing that vessel in the Eta Circle during specific celestial events to collect data and send a new message. It is a repeat-back message versus an original message.

A reasonable outline for creating a repeatable experiment to test the hypothesis might look something like this:

1. Develop a prospectus identifying the project, its justification, its goals, the expected results, and potential benefits for humanity.
2. Identify and assemble the appropriate professional team to execute the project.
3. Identify and agree upon the repeat-back message consistent with the option presented in *Angel Communication Code*, with the understanding that there could be variations considered.
4. Identify the equipment and what sort of oceangoing vessel would be required.
5. Identify a time line for the project from start to finish.
6. Develop a detailed project budget.
7. Identify a source or sources of funding for a project of this nature.

How much would an experiment like this cost?

Figuring out a budget for an experiment like this would start by securing the vessel. The Florida Institute of Oceanography would be a logical place to start based on the location and the scientific personnel who work at the university. They also have a suitable vessel (*RV Weatherbird*) that costs approximately ten thousand dollars per day to charter. It comes well equipped to support a science party of six for twelve days around the clock. The vessel would need to be outfitted with radio and visual telescopes and other project-specific gear. A detailed cost analysis would need to be developed, but it is reasonable to expect that a $2 million–$3 million budget would get the job done. That is short money in the ET communication search game as we have discussed.

To the best of our knowledge, an oceangoing ET search expedition involving numerous scientific disciplines all in the same place at the same time and, most importantly, for the same singular purpose has never been executed. Relatively speaking, it is not a difficult undertaking. It is certainly not as difficult or costly as the launching of satellite messages, construction and continuous operation of space telescopes, and continuing to invest millions and millions of dollars every year—and getting no ET communication results.

Conclusion

The information presented in *Angel Communication Code* is obviously controversial. That is unavoidable as it is the nature of the scientific thought process, exploration, and discovery. History is full of controversial hypotheses that were put to the test and later proven true. According to the *Merriam-Webster Dictionary*, to explore is to "conduct a systematic search or to travel over new territory for adventure or discovery." Throughout history, humankind has shared an innate trait—the desire to explore.

In 1492, the European age of exploration began when the king and queen of Spain financed a voyage by Italian mariner Christopher Columbus. His expedition was to sail west from Europe seeking a more efficient route to India. His commitment to venture into the perils of the unknown has been shared by explorers throughout history. Columbus insightfully once said, "You can never cross the ocean unless you have the courage to lose sight of the shore."

Columbus set out to find faster trade routes to India and the result was the European discovery of North America. His courage and willingness to do what he did resulted in the discovery of a "new world," which was not part of the expedition's expectations, but it ended up changing the world forever. Every time we have gone to the frontier, we have brought back much more than we ever anticipated or even imagined.[1]

This is also the case with our search for intelligent life in the universe over the past many decades. The primary objective has not been achieved; however, hundreds of other discoveries have been made that are very important to our understanding of the universe.

Exploration requires a vision first, then a plan, then financial backing, and finally—and most important—courage. It requires courage to both present the vision and then execute the plan. It took a lot of forward thinking to present the information contained in *Angel Communication Code* because the diversity of information has never been compiled, correlated, linked together, and presented before as a key to ET communication. It is easy to refute the information as coincidental and circumstantial because there is only a theory with no hard proof. We have collected the facts, put all the facts in line to tell the story, and developed the theory. The only proof will be to go to trial, execute the experiment, and then assess the experiment's results. Unless an ET drops out of the sky and strikes up a conversation with humanity, humanity needs to go on an expedition to communicate with them. We have been searching for decades with no results, so it is now time to update the experiment. That is how the system works. It is the scientific method.

The experiment proposed in *Angel Communication Code* is a new vision and a plan for exploration and discovery of a new world. It is an attempt to establish communication with intelligent life beyond Earth.

---

[1] Bob Granath, "The Human Desire for Exploration Leads to Discovery," NASA, October 2, 2015.

It is the exploration of our new frontier. It is different from ET communication efforts that were designed and executed in the past and continue to this day, which have not produced the desired results. That is the very point and justification for what is proposed in *Angel Communication Code*. We are proposing a back-to-basics approach consistent with the tried-and-true scientific method with respect to the development of a valid hypothesis leading to experimental testing of that hypothesis. It is the very definition of exploration: "conduct a systematic search or to travel over new territory for adventure or discovery." As noted, humankind has always shared an innate trait—the desire to explore.

It is our human nature to keep searching to see what is over the next hill, so we need to try a new approach to getting over this particular ET communication hill. Perhaps what we have presented in *Angel Communication Code* is the key to achieving real-time communication with ETs within our lifetimes. We have nothing to lose and everything to gain by trying a different approach as the scientific method demands.

About the Author

Stephen J. Silva's debut book, entitled *Extraterrestrial Communication Code*, was published in 2021. He maintains a supporting website entitled *Extraterrestrial Communication Group* (www.etcommgroup.com) and also a Facebook Page (https://www.facebook.com/etcommgroup/). These online forums are in place to invite open discussions on extraterrestrial communication and related topics.

Stephen Silva is a civil environmental engineer and has more than thirty years of practical engineering experience. His formal education includes a degree in offshore marine technology from the Florida Institute of Technology and a civil environmental engineering degree from the University of Vermont.

For the past decade, he has been conducting research and studying the science of extraterrestrial existence and extraterrestrial communication means and methods. The result of all the research in combination with the dreamlike vision of an extraterrestrial communication code has culminated in the publication of two books on the subject. This book is the second.

These books present information leading directly to a scientifically based hypothesis about establishing ET communication. Furthermore, they present the foundation of an experiment designed to put that hypothesis to the test, consistent with the universally accepted standard scientific method. What is presented in both books is not a topic-by-topic review of what is already known, but rather a pulling together of what is already known and a combining of this information with fresh new insights with a singular focus. These books present original concepts never before pulled together, establishing a common thread leading to a new way to establish two-way ET communication.

Stephen J. Silva is also a proud and faithful Christian who believes in God, which can give rise to a delicate and complex dance when discussing the subject of extraterrestrial communication and the creation of universe. Be assured, however, that personal faith does not bias *Angel Communication Code*'s research, findings or conclusions.

The success of this endeavor is dependent upon the communion and coming together of faith and science-based evidence.

Printed in the United States
by Baker & Taylor Publisher Services